CAMBRIDGE TRACTS IN MATHEMATICS

General Editors

H. BASS, H. HALBERSTAM, J. F. C. KINGMAN

J. E. ROSEBLADE & C. T. C. WALL

---

75. *Topics in ergodic theory*

WILLIAM PARRY

Professor of Mathematics, University of Warwick

Topics in ergodic theory

CAMBRIDGE UNIVERSITY PRESS

CAMBRIDGE

LONDON · NEW YORK · NEW ROCHELLE

MELBOURNE · SYDNEY

PUBLISHED BY THE PRESS SYNDICATE OF THE UNIVERSITY OF CAMBRIDGE
The Pitt Building, Trumpington Street, Cambridge, United Kingdom

CAMBRIDGE UNIVERSITY PRESS
The Edinburgh Building, Cambridge CB2 2RU, UK
40 West 20th Street, New York NY 10011–4211, USA
477 Williamstown Road, Port Melbourne, VIC 3207, Australia
Ruiz de Alarcón 13, 28014 Madrid, Spain
Dock House, The Waterfront, Cape Town 8001, South Africa

http://www.cambridge.org

First published 1981
First paperback edition 2004

A catalogue record for this book is available from the British Library

Library of Congress cataloguing in publication data

Parry, William, 1934-
Topics in ergodic theory.
(Cambridge tracts in mathematics; 75)
Bibliography :p.
Includes index.
1. Ergodic theory. I. Title. II. Series.
QA 313.P37   515'.42   79–7815

ISBN 0 521 22986 3 hardback
ISBN 0 521 60490 7 paperback

To Yael

Contents

Preface

Ergodic theory is difficult to characterise, as it stands at the junction of so many areas, drawing on the techniques and examples of probability theory, vector fields on manifolds, group actions on homogeneous spaces, number theory, statistical mechanics, etc. A comprehensive account of the subject today would necessarily lack the formal unity which one expects of fields wholly contained within one of the main disciplines of mathematics.

Nevertheless, there are excellent general accounts in Hopf [1], Halmos [1], Friedman [1], Jacobs [1], and Walters [1], which display a unity where there might seem to be none, and the survey by Mackey [1] demonstrates conclusively that the seeming chaos of the subject camouflages a very real order – albeit a complex, organic, rather than mechanical one. Of course most books and surveys on ergodic theory reflect their authors' special interests as does the present volume.

The main reason for my writing this monograph is a simple desire to put together some of my favourite topics. I believe, however, that it might serve more than a personal whim. There are many directions a researcher in ergodic theory might take and the chapters in this book could provide the first steps for each of these various journeys. Following the main body of this treatise I have included an appendix on the spectral multiplicity theory of unitary operators. This appendix should help students in so far as the material presented is either left out of most texts or is imbedded in treatments which are unnecessarily exhaustive for our purposes.

I have aimed neither for the utmost generality in the theorems presented nor for scholarly comprehensiveness in my acknowledgement of authorship or of modern trends; in this connection I must beg the indulgence of mathematicians whose contributions, though relevant, have been glossed over or not mentioned at all.

The introduction includes a brief account of the origins of ergodic theory and an outline of present trends. It is included merely to give the reader some feeling for the place of this selection in a rapidly developing and important subject and in the hope that the coherence of history will compensate for the subjectivity of my choice. For some readers it may be advisable to pass over this section until some familiarity with ergodic problems is gained.

The author wishes to acknowledge the critical help received from I. Namioka during the delivery of the lectures on which this work is based, and from Peter Walters, Ralf Spatzier and Selim Tuncel, at a later stage. Dr F. Smithies was especially helpful in highlighting many typographical errors and stylistic blunders. My deepest gratitude is due to Keith Wilkinson for his painstaking sub-editorial assistance and to Heini Halberstam for his interest and encouragement.

Reference symbols Lemmas and theorems are labelled successively 1, 2, ... in each chapter; statements for reference are labelled (1.1), (1.2), ... in Chapter 1 with a similar system for other chapters; in each chapter exercises are numbered successively 1, 2, ... References to books and papers are given by name and a number.

University of Warwick   WILLIAM PARRY
December 1980

0
Introduction

1 Origins

Statistical mechanics as developed by Gibbs [1], grew out of an attempt to apply systematically probability theory (not yet itself systematised) to conservative mechanical systems with many degrees of freedom. Earlier investigations along these lines by Maxwell and Boltzmann were based on highly specialised assumptions concerning the interaction of particles for the purpose of explicating thermodynamics. Gibbs sought a theory which made as few assumptions as possible concerning the 'nature' of particles yet which, at least by analogy and perhaps by direct application, embraced thermodynamical problems. His theory rested on Hamiltonian formalism and thus included an invariant (Liouville) distribution of phase.

If a system of k particles moves in a region of three-dimensional space, at each point of time the system is described by the position $q^i = (q^i_1, q^i_2, q^i_3)$ and momentum $p^i = (p^i_1, p^i_2, p^i_3)$ coordinates for $i = 1$, $2, \ldots, k$. These data are summarised by a point x in some region X of $6k$-dimensional space, the phase space. Assuming the system to be conservative, Liouville's theorem asserts the existence of a volume m on X 'smoothly' related to the differential structure of X, which is invariant through the course of time, i.e. a subregion will evolve through successive subregions of the same volume. If x_t $(t \in \mathbb{R})$ represents the time evolution of a point in phase space, then $T_s: x_t \rightarrow x_{t+s}$ $(s \in \mathbb{R})$ defines a one-parameter group of transformations $(T_t \circ T_s = T_{t+s}, T_0 = \text{identity})$ which preserves the measure m.

If F is a sufficiently smooth real-valued function which is invariant with respect to the motion defined by $\{T_t\}$ $(F \circ T_t = F$ for all $t \in \mathbb{R})$, then the study of the dynamical system on X reduces to the study of the dynamical system on the invariant subspaces $F_a = \{x : F(x) = a\}$ and each F_a possesses an invariant volume m_a canonically related to m. For a Hamiltonian system the Hamiltonian function H (or total energy) is such a function or *integral* and, in general, m_a is a finite invariant measure.

Modern ergodic theory grew out of the so-called ergodic problem or hypothesis, viz. phase averages and time averages coincide on each surface of constant energy:

$$\frac{1}{m_a(H_a)} \int_{H_a} f \, dm_a = \lim_{T \to \infty} \frac{1}{2T} \int_{-T}^{T} f \circ T_t \, dt. \qquad (0.1)$$

Boltzmann thought that each surface of constant energy might be entirely filled by a single trajectory – a hypothesis which would ensure (0.1), but which is rarely satisfied. The quasi-ergodic hypothesis (of P. and T. Ehrenfest) that each trajectory is dense in an energy surface is not enough to ensure (0.1) and in any case excludes numerous important examples.

We now know that (0.1) holds for almost all $x \in H_a$ (or in the $L^2(H_a)$ sense of convergence) if and only if $\{T_t : t \in \mathbb{R}\}$ is *ergodic* on (H_a, m_a) (a term to be defined later). These statements, which amount to Birkhoff's [1] and von Neumann's [1] ergodic theorems respectively, ushered into mathematics the subject of ergodic theory proper. Both theorems were initially formulated for conservative Hamiltonian systems: in other words in the setting of ordinary differential equations, and involved difficulties which have subsequently been removed. von Neumann's theorem was based on Koopman's observation that these mechanical systems, preserving as they do a natural measure, induce on the Hilbert space of square integrable functions a one-parameter group of unitary operators. Thus, von Neumann was able to exploit the spectral theory in the development of which he had played such a major role. Subsequently Hopf provided a simplified proof which avoided spectral theory.

Even though von Neumann's theorem is intrinsically on a more elementary level than Birkhoff's, it was all that was needed (given ergodicity) for the original purposes of statistical mechanics as von Neumann argued and as Birkhoff (and Koopman) partially conceded: 'In view of these facts, it is of interest to decide which of the two formulations, (1) or (2) [mean convergence or a.e. convergence] corresponds to the actual physical problem of the ergodic hypothesis. It turns out that (1) is sufficient – that it, indeed, is the precise mathematical equivalent of the physical state of affairs.' (von Neumann [2], p. 275.)

'With regard to the scope of this theorem, we may make the following remarks:

1. From the point of view of the gross statistics on Ω (classical kinetic theory), it is equivalent in its implications to the Mean Ergodic Theorem.

2. From the viewpoint of the detailed statistics along an individual path curve, it is fundamentally more far-reaching; in it is proved for the first time that the relative time of sojourn along almost every individual path curve *exists*, a result often assumed implicitly in the writing of physicists, but never proved.' (Birkhoff and Koopman [1], p. 281.)

The above reference to time averages on (almost all) trajectories is not without force. Birkhoff's theorem allows us to speak of these averages, in connection with the duration spent by a point in a particular region, and we may conclude, almost surely (when the system is ergodic), that this average time is proportional to the volume of the region. In other words, not only do we know that almost all points of a region return infinitely often to that region, but we can state the proportion of time spent in that region. In this connection the former non-quantitative version of recurrence was known to Poincaré [1] and had been 'proved' by Gibbs [1] (Chapter XII). Clearly then, ergodic-type questions had been examined considerably earlier than by Birkhoff and von Neumann. Indeed, in a rather different way (although not entirely, considering its relation to the problem of Lagrange (cf. Arnold and Avez [1]; Sternberg [1])), we may consider Weyl's theorem on the uniform distribution mod 1 of $n\alpha_1, \ldots, n\alpha_k$ (when $\alpha_1, \ldots, \alpha_k, 1$ are rationally independent), as the first special ergodic theorem and we may note that it bears the same relationship to Kronecker's theorem on the density of this sequence, as does Birkhoff's theorem to Poincaré's recurrence theorem.

Following the publication of the two ergodic theorems a dichotomy in the development of dynamical systems can be discerned. On the one hand investigations were carried out into the topology of dynamical systems – topological dynamics (Gottschalk and Hedlund

[1]; Krylov and Bogolioubov [1]; Ellis [1]) and the qualitative theory of differential equations (Nemytskii and Stepanov [1]) – programmes initiated by Birkhoff [2] and Poincaré [1], respectively – and on the other, the theory of measure-preserving transformations or ergodic theory was pursued abstractly. The two branches proceeded on more or less independent paths.

Ergodic theory itself shot off in various directions. The basic ergodic theorems were clarified by Khintchine [1]; Hopf [1]; Kakutani and Yosida [1]; and new ergodic theorems were proved by Hopf [2], [3]; Hurewicz [1]; Chacon and Ornstein [1] (cf. also Garsia [1], [2]). The problem of the existence of finite and σ-finite invariant measures was pursued (Hopf [4]; Dowker [1], [2]; Hajian and Kakutani [1]; Ornstein [1]), and dynamical systems were decomposed and represented (Ambrose [1]; Halmos [2]; Ambrose, Halmos and Kakutani [1]; Ambrose and Kakutani [1]; Rohlin [1]). As these results are not related to the main themes of this work, we refer the reader to the addresses and surveys of Halmos [3]; Kakutani [1] and Rohlin [2].

One result in particular calls for special emphasis as it has played such an important role in motivating the research of the past three decades. I am referring to the classification theory of Halmos and von Neumann (von Neumann [3]; Halmos and von Neumann [1]) for ergodic measure-preserving transformations with discrete spectrum. Two such transformations on (Lebesgue) probability spaces are *isomorphic* (cf. Chapter 4) if and only if they have the same set of eigenvalues (in this case, are spectrally equivalent). Halmos and von Neumann displayed all such transformations as compact abelian group translations. Very little further progress on the classification problem was made until 1958, although von Neumann and Anzai [1] constructed interesting non-isomorphic but spectrally equivalent transformations with mixed spectrum.

In 1954 Kolmogorov [1] directed the attention of mathematicians to various unsolved problems in classical mechanics connected with Hamiltonian systems and the decomposition of these systems into invariant tori. Concerning these tori he enquired after the possibility of changing flows into the well-known 'irrational' flows by an analytic change of coordinates. These problems involve questions of approximability of irrational numbers by rationals (the so-called problem of 'small denominators') and are allied to problems of structural stability. Further progress on these questions was made by Kolmogorov [2]; Arnold [1], [2], [3], [4]; Moser [1]; Herman [1]. The concept of structural stability just referred to is due to Andronov and Pontrjagin [1] and concerns the problem of what basic features of a dynamical system remain intact when the coefficients of that system undergo a small perturbation. If the system remains qualitatively the same, it is said to be structurally stable. Investigations into this form of stability were revived by the works of Peixoto [1]; Arnold and Sinai [1]. Indeed it was principally these works which led to a mushrooming of activity in differentiable dynamical systems and classical mechanics in recent years (cf. Smale [1] [2], Anosov [1]).

These problems are not central to ergodic theory as such, but the direction of this research is particularly interesting to us as it is possible to discern a return to the kind of problems which gave rise to ergodic theory. Moreover, the impetus which ergodic theory received in 1958 moved it in a direction which relates increasingly to the classical mechanical problems. Thus, the divergent strands of dynamical systems are beginning to converge again.

In 1958 Kolmogorov [2] introduced the concept of entropy (borrowed from Shannon's information theory [1]) into ergodic theory and thereby solved an outstanding problem: to find two spectrally equivalent dynamical systems with *continuous* spectrum which are not (spatially) isomorphic. In fact entropy is an isomorphism invariant assigning easily computable numbers to Bernoulli systems (independent processes – cf. Chapter 3) and all Bernoulli systems have the same continuous spectral characteristics. Since that time ergodic theory has made enormous progress in the hands of Rohlin [3], [4], [5], and Sinai [1], [2], [3], [4], initially, and latterly with Ornstein's outstanding results (Ornstein [2], [3]). Sinai employed the new concept in his analyses of classical dynamical systems [2] and, in particular, in his investigations of hard spherical gases [3]. More recently Ornstein classified *all* Bernoulli systems [3] and, together with Weiss (Ornstein and Weiss [1]), showed that the geodesic flows on surfaces of constant negative curvature are Bernoulli flows. The research of recent years is so abundant that we can do no better than refer the reader to the books and surveys – Shields [1]; Ornstein [2]; Friedman and Ornstein [1]; Weiss [1]; Moser, Phillips and Varadhan [1].

These, then, are some of the main trends in present day ergodic theory. In the early days there were several stages of abstraction implicit in the development of the subject: thermodynamics, statistical mechanics, conservative mechanical systems, measure-preserving transformations. Today, with its new connections with differentiable dynamical systems and aided by the relatively new ideas from information theory, ergodic theory continues to fulfil its earlier promises.

2 Preliminaries

A *probability space* $(X, \mathcal{B}, m)$ is a triple with X a set, $\mathcal{B}$ a σ-algebra of subsets of X, and m a probability measure (a non-negative countably additive function defined on $\mathcal{B}$ with $m(X) = 1$).

A *measure-preserving transformation* (*or endomorphism*) T of $(X, \mathcal{B}, m)$ is a surjective map $T : X \to X$ such that $T^{-1} \mathcal{B} \subset \mathcal{B}$ (i.e. $T^{-1} B \in \mathcal{B}$ whenever $B \in \mathcal{B}$) and $m(T^{-1} B) = m(B)$ for all $B \in \mathcal{B}$.

A measure-preserving transformation T is said to be *invertible* (or an *automorphism*) if T is one – one and $T^{-1} \mathcal{B} = \mathcal{B}$ so that the inverse T^{-1} is also a measure-preserving transformation.

A *measure-preserving flow* is a one-parameter group of measure-preserving transformations $\{T_t : t \in R\}$ (each T_t $(t \in R)$ is measure-preserving, and $T_t \circ T_s = T_{t+s}$, $T_0 = $ identity) such that the map $X \times \mathbb{R} \to X, (x, t) \to T_t x$ is measurable.

As we have said in § 1, measure-preserving flows arise naturally in the study of conservative Hamiltonian dynamical systems. Their investigation can be facilitated by a further reduction – to single measure-preserving transformations – on at least two counts:

(1) If the dynamical system or measure-preserving flow possesses a *global cross-section* as illustrated in the diagram below, then associated with that flow we have a measure-preserving transformation of the cross-section and its study will provide considerable information about the total flow.

A *global cross-section* C has the following property: each orbit (in-

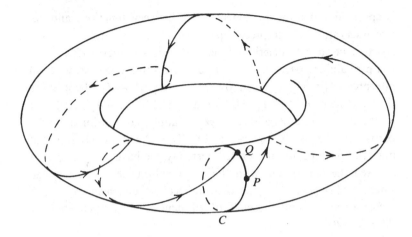

dicated by arrowed lines) passes through C infinitely many times and leaves C immediately. We then have a transformation S (the Poincaré map) of C which maps P to the point Q of first return. (There is then a measure on C, which is canonically associated with the original measure on the total space, which is preserved by S.) Detailed knowledge of S can then be used in an analysis of the original dynamical system.

(2) If we choose $\varepsilon > 0$ very small, then we can expect the 'discrete' flow $\{T_{n\varepsilon} : n = 0, \pm 1, \ldots\}$ to be a good approximation to the flow $\{T_t : t \in \mathbb{R}\}$, and thus it is worthwhile studying the single transformation T_ε and its iterates $T_{n\varepsilon} = T_\varepsilon^n$ where, inductively, $T_\varepsilon^n = T_\varepsilon \circ T_\varepsilon^{n-1}$ ($T_\varepsilon^0 = $ identity).

Except for an example in Chapter 5 and some exercises, we shall be concerned exclusively with single transformations (and their iterates) rather than with measure-preserving flows.

3 Conventions

Throughout this work we shall adopt the following conventions concerning relationships modulo sets of measure zero:

Where several measures are involved, such as in § 1, Chapter 1, equations between functions, especially continuous functions, have to be interpreted strictly. For most of this work, where only one measure is specified, equations and inequalities between functions and between sets are to be interpreted up to a set of measure zero (i.e. they are strictly true after the deletion of some set of measure zero). If $(X, \mathcal{B}, m)$ is a probability space and $\mathcal{A}_1$, $\mathcal{A}_2$ are two sub-σ-algebras, we shall interpret $\mathcal{A}_1 \subset \mathcal{A}_2$ to mean that, for every $A_1 \in \mathcal{A}_1$, there exists $A_2 \in \mathcal{A}_2$ with $m(A_1 \triangle A_2) = 0$. $(A_1 \triangle A_2 = A_1 \cup A_2 - A_1 \cap A_2.)$ $\mathcal{A}_1 = \mathcal{A}_2$ means $\mathcal{A}_1 \subset \mathcal{A}_2$ and $\mathcal{A}_2 \subset \mathcal{A}_1$. A similar convention holds for partitions α_1, α_2 (see Chapters 2 and 4). To make absolutely sure that our relationships are interpreted correctly, we have in many places stressed the convention with the abbreviation a.e. (almost everywhere) or (a.e.) [m] when we wish to specify the measure m.

For a space X and subset B, B^c will denote the complement of B $(B^c = X - B)$.

For a vector space V and subspaces V_1, V_2, $V_1 + V_2$ will denote the subspace consisting of vectors $v_1 + v_2$ with $v_1 \in V_1$, $v_2 \in V_2$. We reserve the convention $V_1 \oplus V_2$ for the case when V_1, V_2 are mutually orthogonal subspaces of a Hilbert space. If $V = V_1 \oplus V_2$, then $V_1^\perp = V_2 = V \ominus V_1$.

1
The principal ergodic theorems

---

1 Uniform distribution (mod 1) and some topological dynamics

We begin with a result of H. Weyl [1] which may be regarded as the first ergodic theorem to be discovered. Admittedly the result is special. It asserts that the sequence $\{x_n\}$ $(x_n = n\alpha)$ is *uniformly distributed* (*mod* 1) when α is irrational, i.e.

$$\frac{1}{N} \sum_{n=0}^{N-1} \chi_I((x_n)) \to |I| \quad \text{for all intervals } I \subset [0, 1], \tag{1.1}$$

where $|I|$ denotes the length of the interval I, ( ) denotes the fractional part of a number, and χ_I denotes the indicator function of I. Weyl gave the following criterion for uniform distribution (mod 1):

$$\text{for each integer } k \neq 0, \ \frac{1}{N} \sum_{n=0}^{N-1} \exp(2\pi i k x_n) \to 0. \tag{1.2}$$

In fact it is easy to see that if (1.1) holds then

$$\frac{1}{N} \sum_{n=0}^{N-1} f(x_n) \to \int_0^1 f(y)\,dy \tag{1.3}$$

for all continuous functions f on $[0, 1]$ with $f(0) = f(1)$, since (1.3) will hold for step functions (by taking linear combinations of characteristic functions) and uniform approximation will lead to (1.3). The condition (1.3) will hold for complex functions when it holds for real functions. Hence, as a special case, (1.2) follows from (1.1) (and (1.3)). Conversely, if (1.2) holds, then by taking linear combinations we have (1.3). Here we use Weierstrass's approximation theorem. It is now easy to deduce (1.1) from (1.3). For $I \subset [0, 1]$ choose continuous real-valued functions f, g with $f \geq \chi_I \geq g$ such that

$$f(0) = f(1), \quad g(0) = g(1), \quad \int_0^1 (f - g)\,dy < \varepsilon.$$

The statement (1.3) for f and g (depending on $\varepsilon > 0$) leads quickly to (1.1). We have shown that:

(1.1), (1.2), (1.3) *are pairwise equivalent.*

Using (1.2) we now see that $\{n\alpha\}$, and more generally $\{n\alpha + x\}$, is uniformly distributed (mod 1) when α is irrational. In fact for $k \neq 0$

$$\frac{1}{N} \sum_{n=0}^{N-1} \exp\left[2\pi ik(n\alpha + x)\right] = \exp(2\pi ikx) \cdot \frac{1}{N}\left[\frac{\exp(2\pi iN\alpha) - 1}{\exp(2\pi i\alpha) - 1}\right] \to 0.$$

We can place these facts into the context of topological dynamics as follows. Define $Tx = x + \alpha$ (mod 1) on $[0, 1]$ with 0, 1 identified. Evidently T is a homeomorphism of a circle and $T^n x = x + n\alpha$ (mod 1).

In this way we see, through the equivalence of (1.2) and (1.3), that

$$\frac{1}{N} \sum_{n=0}^{N-1} f(T^n x) \to \int_0^1 f(y)\,dy \tag{1.4}$$

for all continuous functions defined on the circle. It is clear that T has an equivalent representation as a homeomorphism of the circle $K = \{z \in \mathbb{C} : |z| = 1\}$ given by $Tz = e^{2\pi i\alpha}z$.

In higher dimensions we define $Tx = x + \alpha$ (mod 1), with $x = (x_1, \ldots, x_k)$, $x_i \in [0, 1]$, $\alpha = (\alpha_1, \ldots, \alpha_k)$ and again we identify 0 with 1 in each coordinate. T is a homeomorphism of the k-dimensional torus. (Equivalently, $T(z_1, z_2, \ldots, z_k) = (e^{2\pi i\alpha_1}z_1, \ldots, e^{2\pi i\alpha_k}z_k)$ on the torus $K \times \cdots \times K$ (k times).)

As before (the proof is the same) the following statements are equivalent:

$$\frac{1}{N} \sum_{n=0}^{N-1} \chi_I(T^n x) \to |I| \tag{1.5}$$

for all rectangles $I = I_1 \times \cdots \times I_k$, where $|I|$ denotes the volume of I;

$$\frac{1}{N} \sum_{n=0}^{N-1} \exp 2\pi i \langle h, T^n x \rangle \to 0 \tag{1.6}$$

when $h = (h_1, \ldots, h_k) \neq (0, \ldots, 0)$ (a lattice point), and where $\langle h, x \rangle = h_1 x_1 + \cdots + h_k x_k$;

$$\frac{1}{N} \sum_{n=0}^{N-1} f(T^n x) \to \int_0^1 \cdots \int_0^1 f(y_1, \ldots, y_k)\,dy_1 \cdots dy_k \tag{1.7}$$

for all continuous functions f on the torus, i.e. functions f continuous on $[0, 1] \times \cdots \times [0, 1]$ with $f(0, x_2, \ldots, x_k) = f(1, x_2, \ldots, x_k)$ etc.

We conclude that (1.5), (1.6), (1.7) hold when $\alpha_1, \ldots, \alpha_k$, 1 are rationally independent; for then $\langle h, \alpha \rangle$ is never an integer (h a lattice point) except when $h = (0, \ldots, 0)$ and hence, when $h \neq (0, \ldots, 0)$,

$$\frac{1}{N} \sum_{n=0}^{N-1} \exp 2\pi i \langle h, T^n x \rangle$$

$$= \frac{1}{N} \sum_{n=0}^{N-1} \exp 2\pi i \langle h, x + n\alpha \rangle$$

$$= \frac{1}{N} \exp 2\pi i \langle h, x \rangle \frac{(\exp 2\pi i N \langle h, \alpha \rangle - 1)}{(\exp 2\pi i \langle h, \alpha \rangle - 1)} \to 0.$$

The above results are due to Weyl [1] who also proved:

If $p(\lambda) = a_0 + a_1\lambda + \cdots + a_k\lambda^k$ is a polynomial with real coefficients, with one of $a_1, \ldots, a_k$ irrational, then $\{p(n) : n = 1, 2, \ldots\}$ is uniformly distributed (mod 1).

It is not so obvious how to put this result into the framework of topological dynamics. Nevertheless, this has been done by Furstenberg [1] and Hahn [1]. We shall indicate the main idea by showing that $p(n) = a_0 + a_1 n + a_2 n^2$ is uniformly distributed (mod 1) when a_2 is irrational (cf. Exercise 14). Let α be irrational and define on the 2-dimensional torus the following 'affine' transformation:

$$T(x_1, x_2) = (\alpha + x_1, x_1 + x_2) \quad (\text{mod } 1).$$

Evidently,

$$T^n(x_1, x_2) = (n\alpha + x_1, \tfrac{1}{2}n(n-1)\alpha + nx_1 + x_2) \quad (\text{mod } 1).$$

Now let $\tfrac{1}{2}\alpha = a_2$, $x_1 - \tfrac{1}{2}\alpha = a_1$, $x_2 = a_0$ above. It will suffice to prove (Exercise 14) that

$$\frac{1}{N} \sum_{n=0}^{N-1} f(T^n x) \to \int_0^1 \int_0^1 f(y_1, y_2) \, dy_1 \, dy_2 \tag{1.8}$$

for all continuous functions f on the 2-torus. For then, taking $f(y_1, y_2) = \exp 2\pi i k y_2$ ($k \neq 0$), we shall have

$$\frac{1}{N} \sum_{n=0}^{N-1} \exp 2\pi i k p(n) \to 0.$$

The convergence in (1.7) and (1.8) for the homeomorphisms specified is actually uniform. For (1.7) this can be proved directly. In fact, there is a general result here which amounts to a topological ergodic theorem. The following results are mainly due to Krylov and Bogolioubov [1] (cf. also Oxtoby [1]; Furstenberg [1]).

Let T be a homeomorphism of a compact metric space X and let $C(X)$ denote the Banach space of complex-valued continuous functions on X. The closed subspace of T-invariant functions $f(f = f \circ T)$ will be denoted by I and B will denote the subspace of functions of the form $f - f \circ T$. (In general B is not closed.) From now on we refer to expressions like (1.7) or (1.8) as *time averages*.

1. Theorem *The following are equivalent*:

(i) $(1/N)\sum_{n=0}^{N-1} f(T^n x) \to \tilde{f}(x)$ *for all* x *and each* f, $f \in C(X)$, *where* $\tilde{f} \in C(X)$ *depends on* f.

(ii) *As* (i) *but the convergence is uniform.*

(iii) $C(X) = I + \bar{B}$.

Proof We first note that for $f \in I$ the uniform convergence of $(1/N)\sum_{n=0}^{N-1} f \circ T^n \equiv f$ is trivial.

Almost as trivial is the uniform convergence of $(1/N) \cdot \sum_{n-1}^{N-1} f \circ T^n$ to 0, when $f = g \circ T - g \in B$, or by approximation, when $f \in \bar{B}$. Clearly $\bar{B} \cap I = \{0\}$, since time averages of functions in I leave the functions unaltered and time averages of functions in $\bar{B}$ are zero. These remarks also show that (iii)$\Rightarrow$(ii). (ii)$\Rightarrow$(i) immediately. We proceed now to the proof that (i)$\Rightarrow$(iii).

Assuming (i) we let

$$P(f) = \lim_{N \to \infty} \frac{1}{N} \sum_{n=0}^{N-1} f \circ T^n = \tilde{f} \in I$$

and note that $f = f - Pf + Pf$ where $f - Pf \in \ker P = \{g \in C(X) : Pg = 0\}$. It will suffice to show that $\ker P = \bar{B}$. By the remarks above, $\ker P \supset \bar{B}$ is closed since P is continuous. Hence we need only demonstrate that a continuous linear functional J annihilates $\ker P$ when it annihilates $\bar{B}$ (or equivalently, B). In fact, if $J(f - f \circ T) = 0$ for all $f \in C(X)$, then, using Lebesgue's dominated convergence theorem for continuous linear functionals, we have for $f \in \ker P$

$$\frac{1}{N} \sum_{n=0}^{N-1} f \circ T^n \to 0,$$

$$\frac{1}{N} \sum_{n=0}^{N-1} J(f \circ T^n) \to 0,$$

$$\frac{1}{N} \sum_{n=0}^{N-1} J(f) \to 0.$$

In other words, $J(f) = 0$ for all $f \in \ker P$ and the theorem is proved.

The dominated convergence theorem we have used above is the following: If f_n, $f \in C(X)$ and $|f_n| \leq g \in C(X)$, then $f_n(x) \to f(x)$ for all $x \in X$ implies $Jf_n \to Jf$ for any continuous linear functional J. This can be seen from:

Riesz representation theorem (Dunford and Schwartz [1])   *If J is a positive linear functional on $C(X)$ with $J(1) = 1$, then there is a unique Borel probability m such that $J(f) = \int f \, dm$ for all $f \in C(X)$.*

Decomposition of functionals (Dunford and Schwartz [1])   *Every continuous linear functional J on $C(X)$ can be written $J = J_1 - J_2 + i(J_3 - J_4)$ where J_1, J_2, J_3, J_4 are positive linear functionals.*

Lebesgue's dominated convergence theorem, together with these, proves the dominated convergence theorem we have used.

Exercise 1   Given a homeomorphism T of a compact metric space X and a probability m defined on the Borel sets of X, show that $J(f \circ T) = J(f)$, where $J(f) = \int f \, dm$ ($f \in C(X)$), if and only if T is measure-preserving.

Exercise 2   Given a homeomorphism T of a compact metric space X, we have a corresponding map $\mu \to \mu T^{-1}$ on the compact convex set P of probabilities in the dual space of $C(X)$ with the weak*-topology. Using the Schauder – Tihonov fixed point theorem or the Markov – Kakutani fixed point theorem (Dunford and Schwartz [1]) show that there is a T-invariant probability defined on the Borel sets of X.

A measure-preserving transformation T of a probability space

$(X, \mathscr{B}, m)$ is said to be *ergodic* if $T^{-1}B = B (B\in\mathscr{B})$ implies $m(B) = 0$ or 1. m is also said to be an *ergodic measure* for T.

Exercise 3 If T is a homeomorphism of the compact metric space X, show that the compact convex set M of T-invariant probabilities (with the weak*-topology) has for its extreme points precisely the ergodic probabilities of T. In particular, conclude that there is always a T-invariant ergodic probability. (Use the Krein – Milman theorem (Dunford and Schwartz [1]).)

We give now a direct proof of the existence of a T-invariant probability for a homeomorphism of a compact metric space X. Since $C(X)$ is separable we can choose a sequence $f_1, f_2, \ldots$ dense in $C(X)$. Let $x\in X$ be any point, fixed for the rest of this construction. We note that $(1/N)\sum_{n=0}^{N-1} f_1(T^n x)$ is a bounded sequence of complex numbers and therefore there is a convergent subsequence involving the integers $N = m_1^1, m_2^1, \ldots$. Let this sequence of integers be $\mathscr{N}_1$. Again for $N\in\mathscr{N}_1$, $(1/N)\sum_{n=0}^{N-1} f_2(T^n x)$ is a bounded sequence of complex numbers so that for some subsequence $\mathscr{N}_2\subset\mathscr{N}_1$,

$$\lim_{\substack{N\to\infty \\ N\in\mathscr{N}_2}} \frac{1}{N}\sum_{n=0}^{N-1} f_2(T^n x) \quad \text{exists.}$$

Repeating this argument indefinitely we obtain sequences of integers $\mathscr{N}_1\supset\mathscr{N}_2\supset\ldots$, where $\mathscr{N}_i = m_1^i, m_2^i, \ldots$, such that $\lim_{N\in Ni} (1/N)\sum_{n=0}^{N-1} f_j(T^n x)$ exists for $j\le i$. Taking the diagonal $\mathscr{N} = m_1^1, m_2^2, \ldots$ we have $\lim_{N\in\mathscr{N}} (1/N)\sum_{n=0}^{N-1} f_i(T^n x)$ exists for all i. Since $\{f_i\}$ is dense in $C(X)$, $\lim_{N\in\mathscr{N}} (1/N)\sum_{n=0}^{N-1} f(T^n x) = J(f)$ exists for all $f\in C(X)$, and defines a linear functional which is clearly positive. Moreover, $J(1) = 1$. Clearly $J(f\circ T) = J(f)$ for all $f\in C(X)$. Writing $J(f) = \int f\, dm$, we see that m is a T-invariant probability.

We are now in a position to prove the following corollary to Theorem 1.

2. Theorem *If T is a homeomorphism of the compact metric space X,* *then there is one and only one T-invariant probability if and only if one of the following holds:*

   (i) *For each* $f \in C(X)$, $(1/N) \sum_{n=0}^{N-1} f(T^n x)$ *converges to a constant (depending on* f*) for all* $x \in X$.
   (ii) *As* (i) *with uniform convergence.*
   (iii) $C(X) = \mathbb{C} + \bar{B}$.

Proof The equivalence of (i), (ii), (iii) is proved in the same way as Theorem 1. We know there is always one T-invariant probability. If m is T-invariant then (i) implies

$$\frac{1}{N} \sum_{n=0}^{N-1} f(T^n x) \to k,$$

where k is a constant depending on f. Using Lebesgue's dominated convergence we have

$$\int f \, dm = \frac{1}{N} \sum_{n=0}^{N-1} \int f(T^n x) \, dm \to \int k \, dm = k,$$

i.e. $\int f \, dm$ is determined *uniquely*.

On the other hand, if (i) is untrue, then there exists $f \in C(X)$ with $(1/N) \sum_{n=0}^{N-1} f(T^n x)$ not converging to a constant. Either there exists x with $(1/N) \sum_{n=0}^{N-1} f(T^n x)$ not converging, in which case

$$\lim_{N \in \mathcal{N}'} \frac{1}{N} \sum_{n=0}^{N-1} f(T^n x) \neq \lim_{N \in \mathcal{N}''} \frac{1}{N} \sum_{n=0}^{N-1} f(T^n x)$$

for different sequences $\mathcal{N}'$, $\mathcal{N}''$, or

$$\lim \frac{1}{N} \sum_{n=0}^{N-1} f(T^n x) \neq \lim \frac{1}{N} \sum_{n=0}^{N-1} f(T^n y)$$

for two different points x, y.

In either case, the construction we have outlined will produce distinct T-invariant measures m_1, m_2 with $\int f \, dm_1 \neq \int f \, dm_2$. In other words, there are at least two T-invariant probabilities, and the corollary is proved.

The phenomenon described in Theorem 2 has a name: a homeomorphism T of a compact metric space is called *uniquely ergodic* if there is only one T-invariant Borel probability.

Notice that if T is uniquely ergodic (with invariant probability m) then T is ergodic with respect to m. This is proved by one of the exercises but, more directly, should $T^{-1}B = B$ with $1 > m(B) > 0$ then

$$m_\lambda(A) = \frac{\lambda m(A \cap B)}{m(B)} + \frac{(1-\lambda)m(A \cap B^c)}{m(B^c)}$$

is a continuum of T-invariant probabilities $(0 \le \lambda \le 1)$.

A homeomorphism T of a compact metric space X is said to be *minimal* if $TY = Y(Y$ closed) implies $Y = \emptyset$ or X. (Equivalently, every orbit $O(x) = \{T^n x\}$ is dense in X.) If T is uniquely ergodic with invariant probability m and if $m(U) > 0$ for all non-empty open sets U, then T is minimal. For should $TY = Y$, with Y closed, we could consider the system $T|Y$. This would have an invariant probability which, assigning zero to Y^c, extends to X. Hence this must be *the T-invariant probability and $m(Y^c) = 0$. By hypothesis Y^c must therefore be empty and $Y = X$. Minimality does *not* imply unique ergodicity (or ergodicity) as we shall see (cf. Chapter 5, § 5).

Exercise 4   If T is a homeomorphism of a compact metric space X, show that X has a minimal subset i.e. there exists $Y \subset X$, $TY = Y$, Y closed and $T|Y$ is minimal on Y. (Use Zorn's lemma on the collection of T-invariant closed sets.)

Exercise 5   Let X be the one-point compactification of the integers. Define $Tx = x + 1, x \neq \infty, T\infty = \infty$. Show that T is uniquely ergodic but not minimal. What is the unique invariant probability? Construct a homeomorphism of the circle with a similar behaviour.

Exercise 6   Show that the translation of the torus $T(x_1, \ldots, x_k) = (x_1 + \alpha_1, \ldots, x_k + \alpha_k)$ (mod 1) is minimal, uniquely ergodic or ergodic with respect to Haar (Lebesgue) measure if and only if $\alpha_1, \ldots, \alpha_k, 1$ are rationally independent.

Exercise 7   Analyse the one-parameter group of homeomorphisms T_t, $T_t(x_1, \ldots, x_k) = (x_1 + t\alpha_1, \ldots, x_k + t\alpha_k)$ (mod 1), and show that analogous theorems to those proved in this section for a single transformation hold. Show that the condition '$\alpha_1, \ldots, \alpha_k, 1$ are rationally independent' should be replaced by the condition '$\alpha_1, \ldots, \alpha_k$ are rationally independent'.

Exercise 8   (From Arnold and Avez [1]) Show that the coefficient in the highest term of the decimal expansion of 2^n, as n runs through the integers, is less frequently 9 than any other digit. Compute this frequency.

Exercise 9   For $x \in \mathbb{R}^+$ define $Tx = \frac{1}{2}x$ when $x > 1$, $Tx = 3x$ when $x < 1$. Describe the behaviour of this system. Is there an irrational translation at work somewhere in this system?

Exercise 10   Show that the automorphism of the 2-torus given by $T(x_1, x_2) = (2x_1 + x_2, x_1 + x_2)$ (mod 1) is not uniquely ergodic and not minimal. (Look for fixed and periodic points.)

Exercise 11   Assuming Weyl's theorem on the uniform distribution of $p(n)$ (for certain polynomials p) show that the affine transformation $T(x_1, x_2) = (\alpha + x_1, x_1 + x_2)$ (mod 1) is uniquely ergodic when α is irrational.

Exercise 12   Show that (1.1) is equivalent to (1.3) with the condition $f(0) = f(1)$ omitted, and even with Riemann integrable functions replacing continuous functions. Can we replace Riemann integrable functions with Lebesgue integrable functions?

Exercise 13   If T is a measure-preserving transformation of the probability space $(X, \mathcal{B}, m)$, then T is ergodic if and only if $f \circ T = f$ a.e. (f measurable) implies f is constant a.e.

3. Theorem (Furstenberg [1])   *If T is a uniquely ergodic homeomorphism of a compact metric space X and if $\varphi: X \to K = \{z \in \mathbb{C} : |z| = 1\}$ is continuous, then the homeomorphism $S: X \times K \to X \times K$, $S(x, z) = (Tx, \varphi(x)z)$ is uniquely ergodic if and only if S is ergodic with respect to $m \times l$, where m is the unique measure for T and l is the Lebesgue measure.*
   Proof $m \times l$ is S-invariant. Let μ be any other S-invariant probability. Since $B \to \mu(B \times K)$ is T-invariant, we have, by the

unique ergodicity of T,

$$\int f(x)\,\mathrm{d}m = \int \int f(x)\,\mathrm{d}(m \times l) = \int \int f(x)\,\mathrm{d}\mu \quad \text{for } f \in C(X).$$

Let $H = L^2(X, m)$ and note that $V: f \to f \circ T \cdot \varphi$ is a unitary operator on H and that $J(f) = \int \int f(x)z\,\mathrm{d}\mu$ is a continuous linear functional on H. Moreover

$$J(Vf) = \int \int f(Tx)\varphi(x)z\,\mathrm{d}\mu = \int \int f(x)z\,\mathrm{d}\mu = J(f),$$

since μ is S-invariant. Hence $J(f) = \langle f, g \rangle$ for some $g \in H$ and $\langle Vf, g \rangle = \langle f, g \rangle$ for all $f \in H$, i.e. $Vg = g$. Therefore $g \circ T \cdot \varphi = g$. Hence $(x, z) \to g(x)z$ is S-invariant and, with the use of Exercise 13 and the ergodicity of S (with respect to $m \times l$), g must be trivial ($g = 0$ a.e. $[m]$). We conclude that J is trivial and $\int \int f(x)z\,\mathrm{d}\mu = 0$. The same proof with z^n ($n \neq 0$) replacing z shows that $\int \int f(x)z^n\mathrm{d}\mu = 0$ ($n \neq 0$). Certainly $\int \int f(x)z^n\mathrm{d}(m \times l) = 0$ ($n \neq 0$) and $\int \int f(x)\,\mathrm{d}\mu = \int \int f(x)\,\mathrm{d}(m \times l)$. Since μ and $m \times l$ agree on a spanning subset of $C(X \times K)$, we see that $\mu = m \times l$, i.e. S is uniquely ergodic.

Here is another proof that $T(x, y) = (\alpha + x, \beta + y)$ (mod 1) is ergodic, when α, β, 1 are rationally independent. Let $f \in L^2(K \times K)$ and assume $f \circ T = f$. Let

$$f(x, y) = \sum_{m, n} a_{m, n} \exp 2\pi\mathrm{i}(mx + ny)$$

be its Fourier series. Then

$$\sum_{m, n} a_{m, n} \exp.2\pi\mathrm{i}(mx + ny) \cdot \exp 2\pi\mathrm{i}(m\alpha + n\beta)$$

$$= \sum_{m, n} a_{m, n} \exp 2\pi\mathrm{i}(mx + ny),$$

i.e. $\exp 2\pi\mathrm{i}(m\alpha + n\beta) = 1$ when $a_{m, n} \neq 0$.

We conclude that $a_{m, n} \neq 0$ implies $m = n = 0$. In other words $a_{0, 0}$ is the only non-zero Fourier coefficient and hence f is constant a.e. From Exercise 13 and Theorem 3 we see that T is ergodic and uniquely ergodic.

***Exercise* 14** Show that the transformation of the torus given by $S(x, y) = (\alpha + x, x + y)$ (mod 1) is ergodic with respect to Lebesgue measure when α is irrational. (Represent an S-invariant L^2 function $f(x, y)$ by its Fourier series $f(x, y) = \sum_{m, n} a_{m, n} \exp 2\pi\mathrm{i}(mx + ny)$ and show that it is constant a.e.) Deduce that S is uniquely ergodic, i.e.

(1.8) holds. This is the promised proof that $p(n) = a_0 + a_1 n + a_2 n^2$ is uniformly distributed (mod 1) when a_2 is irrational.

2 Recurrence and the ergodic theorems of von Neumann and Birkhoff

Just as Weyl provided a quantitative version of the distribution of $(n\alpha_1, \ldots, n\alpha_k)$ in contrast to the Kronecker – Dirichlet qualitative description (the density or non-density of this sequence) so the ergodic theorems of von Neumann and Birkhoff provide a quantitative analysis of recurrence. The qualitative fact of recurrence for finite measure-preserving transformations was proved by Poincaré. It had also been noticed by Gibbs [1]. Khintchine gives a supplementary theorem on recurrence.

4. Theorem *Recurrence theorem of Poincaré – Gibbs. If T is a measure-preserving transformation of the probability space $(X, \mathscr{B}, m)$ and $B \in \mathscr{B}$, then almost all points of B return infinitely often to B, i.e.*

$$m \left(B \cap \bigcap_{N=0}^{\infty} \bigcup_{n=N}^{\infty} T^{-n}B \right) = m(B).$$

Proof We write $B_N = \bigcup_{n=N}^{\infty} T^{-n}B$, so that $B_0 \supset B_1 \supset \cdots$ and $T^{-N}B_0 = B_N$. Evidently each B_N has the same measure and the inclusions ensure that

$$B_0 = B_1 = B_2 = \cdots \quad \text{a.e.}$$

Hence

$$\bigcap_{N=0}^{\infty} B_N = B_0 = B_1 = \cdots \quad \text{a.e.}$$

and

$$B \cap \bigcap_{N=0}^{\infty} B_N = B \cap B_0 = B \quad \text{a.e.}$$

The proof is complete.

The next theorem is so transparent nowadays that it is all too easy to underestimate its significance in the history of ergodic theory. The original proof was more complicated, being based on the spectral theory of Stone and von Neumann. The essential, simple, but all-important idea of representing a measure-preserving transformation by the induced unitary operator was due to Koopman [1]. The main simplification in the proof is due to Hopf [1]. One should notice the decomposition of the Hilbert space which is used in the proof – a technique used in the last section and one to be employed in the proof of the much harder theorem of Birkhoff.

Let T be a measure-preserving transformation of the probability space $(X, \mathcal{B}, m)$. Since $\int f \circ T \, dm = \int f \, dm$ for characteristic functions, we see that the same equation is valid for simple functions and hence for all functions $f \in L^1(x, \mathcal{B}, m)$. Hence

$$\int f \circ Tg \circ \overline{T} \, dm = \int f\bar{g} \, dm \quad \text{for all} \quad f, g \in L^2(X, \mathcal{B}, m),$$

and therefore the operator $U_T: f \to f \circ T$ on the Hilbert space $L^2(X)$ is an isometry. If T is *invertible* (so that $T\mathcal{B} = \mathcal{B}$ and $T^{-1} \circ T$ = identity), then U_T has an inverse and hence is unitary. Exercise 13 shows that T is ergodic if and only if the eigenvalue 1 (corresponding to constant eigenfunctions) for U_T is simple.

We shall need some facts concerning the conditional expectation operator. If $\mathcal{A}$ is a sub-σ-algebra of $\mathcal{B}$, then a function $f \in L^1(X, \mathcal{B}, m)$ (in particular if $f \in L^2(X, \mathcal{B}, m)$) defines a complex or real signed measure μ on $\mathcal{A}$ by $\mu(A) = \int_A f \, dm (A \in \mathcal{A})$, which is absolutely continuous with respect to $m|\mathcal{A}$. By the Radon – Nikodym theorem there exists a unique (modulo sets of measure zero) $L^1(X, \mathcal{A}, m)$ function denoted by $E(f|\mathcal{A})$, called the *conditional expectation of* f *given* $\mathcal{A}$, such that $\mu(A) = \int_A E(f|\mathcal{A}) \, dm$ for all $A \in \mathcal{A}$.

$E(f|\mathcal{A})$ is defined uniquely a.e. by

(i) $\int_A E(f|\mathcal{A}) \, dm = \int_A f \, dm$ for all $A \in \mathcal{A}$;

(ii) $E(f|\mathcal{A}) \in L^1(X, \mathcal{A}, m)$.

The operator $E(\ |\mathcal{A})$ enjoys the following properties:

(iii) For all $f \in L^1(X, \mathcal{B}, m)$, $g \in L^\infty(X, \mathcal{A}, m)$,

$$E(fg|\mathcal{A}) = gE(f|\mathcal{A});$$

(iv) E is a linear idempotent operator on $L^1(X, \mathcal{B}, m)$ and $E(E(f|\mathcal{A}_1)|\mathcal{A}_2) = E(f|\mathcal{A}_2)$ when $\mathcal{A}_1 \supset \mathcal{A}_2$;

(v) $|E(f|\mathcal{A})| \le E(|f| \,|\mathcal{A})$ for all $f \in L^1(X, \mathcal{B}, m)$; when $(1/p) + (1/q) = 1$,

$$E(|fg| \,|\mathcal{A}) \le E(|f|^p|\mathcal{A})^{1/p} E(|g|^q|\mathcal{A})^{1/q}$$

for all $f \in L^p(X, \mathcal{B}, m)$ and $g \in L^q(X, \mathcal{B}, m)$.

If T is a measure-preserving transformation, then

(vi) $E(f|\mathscr{A}) \circ T = E(f \circ T | T^{-1}\mathscr{A})$.

The *conditional probability given* $\mathscr{A}$ is denoted by $m(\ |\mathscr{A})$ and $m(B|\mathscr{A}) = E(\chi_B|\mathscr{A})$. Conditional expectation and probability amount to integration and measure when $\mathscr{A} = \mathscr{N}$, the trivial σ-algebra consisting of sets of measure zero and one: $E(f|\mathscr{N}) = \int f \, dm$, $m(B|\mathscr{N}) = m(B)$. On the other hand when $\mathscr{A} = \mathscr{B}$, $E(f|\mathscr{B}) = f$, $m(B|\mathscr{B}) = \chi_B$.

If $f \in L^2(X, \mathscr{B}, m)$, then $E(f|\mathscr{A})$ is precisely the orthogonal projection of f onto $L^2(X, \mathscr{A}, m)$, since $E(f|\mathscr{A}) \in L^2(X, \mathscr{A}, m)$ by (v) and $f = (f - E(f|\mathscr{A})) + E(f|\mathscr{A})$ is the orthogonal decomposition of f into a vector orthogonal to $L^2(X, \mathscr{A}, m)$ and a vector belonging to $L^2(X, \mathscr{A}, m)$. (If $g \in L^\infty(X, \mathscr{A}, m)$, a dense subspace of $L^2(X, \mathscr{A}, m)$, $\langle f - E(f|\mathscr{A}), g \rangle = \int [f\bar{g} - E(f\bar{g}|\mathscr{A})] \, dm = 0$.)

If T is a measure-preserving transformation of $(X, \mathscr{B}, m)$, then for each $1 \leq p \leq \infty$ the subspace I of $L^p(X, \mathscr{B}, m)$ consisting of T-invariant functions is precisely $L^p(X, \mathscr{I}, m)$, where $\mathscr{I} = \{B \in \mathscr{B} : T^{-1}B = B \text{ a.e.}\}$. To see this, notice that functions in I are $\mathscr{I}$ measurable. On the other hand, if $f \in L^p(X, \mathscr{I}, m)$, then $f \circ T \in L^p(X, \mathscr{I}, m)$, and the equation $f = f \circ T$ a.e. is verified by integrating f and $f \circ T$ over each set in $\mathscr{I}$.

5. Theorem *von Neumann's 'mean' ergodic theorem* [1]. *If* U *is an isometry on the Hilbert space* H, *then for all* $x \in H$

$$\frac{1}{N} \sum_{n=0}^{N-1} U^n x \to Px,$$

where P *is the orthogonal projection onto the subspace* I *of* U-*invariant vectors. In particular, if* T *is a measure-preserving transformation of the probability space* $(X, \mathscr{B}, m)$, *then for all* $f \in L^2(X, \mathscr{B}, m)$

$$\frac{1}{N} \sum_{n=0}^{N-1} f \circ T^n \to E(f|\mathscr{I}) \quad (L^2 \text{ convergence})$$

where $\mathscr{I} = \{B \in \mathscr{B} : T^{-1}B = B \text{ a.e.}\}$. *When* T *is ergodic* $E(f|\mathscr{I})$ *is the constant* $\int f \, dm$.

Proof The last conclusion follows immediately from the first part. We shall prove the first part of the theorem under the extra hypothesis that U is unitary (that T is invertible); the general case is left as an exercise.

Let $I = \{x : Ux = x\}$ and let $B = \{Ux - x : x \in H\}$. If $y \perp \bar{B}$ then $\langle Ux, y \rangle = \langle x, y \rangle$ for all $x \in H$. Hence $U^{-1}y - y \perp x$ for all $x \in H$, i.e. $U^{-1}y = y = Uy$. Therefore $\bar{B}^\perp \subset I$. In fact $I \subset \bar{B}^\perp$ since $Ux = x$ implies $\langle x, Uy - y \rangle = 0$ for all $y \in H$. Hence $H = I \oplus \bar{B}$. Writing $x = Px + y$, $y \in \bar{B}$ we see immediately that $(1/N) \sum_{n=0}^{N-1} U^n y \to 0$ and therefore $(1/N) \sum_{n=0}^{N-1} U^n x \to Px$.

Exercise 15   Prove the theorem for an isometry, i.e. do not assume that U is invertible.

6. Theorem *Khintchine's recurrence theorem* [2]. *If T is a measure-preserving transformation of the probability space $(X, \mathscr{B}, m)$ then for every set $B \in \mathscr{B}$ and every $\varepsilon > 0$ there exists a relatively dense sequence $\mathscr{N}$ of positive integers such that*

$$m(T^{-n}B \cap B) \geq m(B)^2 - \varepsilon$$

for all $n \in \mathscr{N}$.

Proof $\mathscr{N}$ is said to be *relatively dense* if there exists $L > 0$ such that $\mathscr{N}$ intersects every interval of integers of length L. Apply von Neumann's theorem to χ_B to get

$$\left\| \frac{1}{N} \sum_{n=0}^{N-1} \chi_B \circ T^n - P(\chi_B) \right\| < \varepsilon$$

for some integer N, then

$$\left\| \frac{1}{N} \sum_{n=0}^{N-1} \chi_B \circ T^{n+l} - P(\chi_B) \right\| < \varepsilon$$

for the same integer N and *all* l, since $P(\chi_B)$ is T-invariant. Hence, putting $S_N = (1/N) \sum_{n=0}^{N-1} \chi_B \circ T^n$, we have

$$\| S_N \circ T^l - S_N \| < 2\varepsilon \quad \text{for all } l,$$

i.e. $2\|S_N\|^2 - 2\langle S_N, S_N \circ T^l \rangle < 2\varepsilon$ for all l. On the other hand,

$$0 \leq \| S_N - m(B) \|^2 = \| S_N \|^2 - m(B)^2.$$

Therefore

$$m(B)^2 < \|S_N\|^2 < \varepsilon + \langle S_N, S_N \circ T^l \rangle$$

for all integers l. But

$$\langle S_N, S_N \circ T^l \rangle = \frac{1}{N^2} \sum_{n,m=0}^{N-1} m(T^{-n-l}B \cap T^{-m}B).$$

Hence there exist $n, m\, (0 \le n, m < N - 1)$ with

$$m(T^{m-n-l}B \cap B) = m(T^{-n-l}B \cap T^{-m}B) > m(B)^2 - \varepsilon.$$

We note that $-N < m - n < N$ so that $-N - l < m - n - l < N - l$, i.e. in every interval of length $2N$ there exists k with $m(T^{-k}B \cap B) > m(B)^2 - \varepsilon$. Here we have assumed $m - n - l \le 0$, which will be the case for large enough l.

We proceed now to Birkhoff's theorem. Until recently proofs of this theorem tended to follow the simplified lines introduced by Khintchine [1], Kakutani and Yosida [1] and especially Riesz [1]. Garsia [1], [2] improved these simplifications as far as the maximal theorem is concerned and introduced further ideas analogous to the decomposition technique we have already seen.

In the following, $E(\ |\mathscr{I})$ will be defined on $L^1(X, \mathscr{B}, m)$.
$(\mathscr{I} = \{B \in \mathscr{B} : T^{-1}B = B\}).$

7. Decomposition lemma *If T is a measure-preserving map of $(X, \mathscr{B}, m)$, then $B = \{f \circ T - f : f \in L^1(X)\}$ is dense in* $\ker E(\ |\mathscr{I})$ $= \{f \in L^1(X) : E(f|\mathscr{I}) = 0\}$. *Consequently,* $L^1(X) = \bar{B} + I$ *and* $\bar{B} \cap I = \{0\}$ *where* $I = \{f \in L^1(X) : f \circ T = f\}$.

Proof The second statement follows from the first since $L^1(X) = \ker E(\ |\mathscr{I}) + I (f = f - E(f|\mathscr{I}) + E(f|\mathscr{I}))$ and $I \cap \ker E(\ |\mathscr{I}) = \{0\}$. Since $E(f \circ T - f|\mathscr{I}) = E(f \circ T|\mathscr{I}) - E(f|\mathscr{I}) = E(f|\mathscr{I}) \circ T - E(f|\mathscr{I}) = 0$ (this follows from $T^{-1}\mathscr{I} = \mathscr{I}$), we see that $B \subset \ker E(\ |\mathscr{I})$. To prove the lemma it will suffice to show that $\int gh \,dm = 0$ for all $g \in B$, where $h \in L^\infty(X)$, implies $\int gh \,dm = 0$ for all $g \in \ker E(\ |\mathscr{I})$, since $L^\infty(X)$ is the dual space of $L^1(X)$. To this end we suppose $h \in L^\infty(X)$ and $\int (f \circ T - f)h \,dm = 0$ for all $f \in L^1(X)$. Since $L^\infty(X) \subset L^1(X)$, we have $\int (h \circ T - h)h \,dm = 0$ and $\int (h \circ T - h)h \circ T \,dm = 0$ (this is an equivalent statement), so that $\int (h \circ T - h)^2 \,dm = 0$, i.e. $h \circ T = h$. We shall show that $\int gh \,dm = 0$ if $E(g|\mathscr{I}) = 0$. In fact $\int gh \,dm = \int E(gh|\mathscr{I})\,dm = \int h E(g|\mathscr{I})\,dm = 0$, and this completes the proof.

Let $B_0 = \{g \circ T - g : g \in L^\infty(X)\}$ and note that $\bar{B}_0 = \bar{B}$. If $f \in B_0 + I$ then $f = f_1 + f_2$, $f_1 \in B_0$, $f_2 \in I$ and $(1/N)\sum_{n=0}^{N-1} f(T^n x) \to f_2(x)$ for almost all $x \in X$. We wish to deduce a.e. pointwise convergence for each $f \in L^1(X)$. For this we need:

8. Maximal ergodic theorem *Let U be a positive contraction on $L^1_{\mathbb{R}}(X)$ (i.e. $\|U\| \le 1$ and $Uf \ge 0$ when $f \ge 0$), and let $f^n = f + Uf + \cdots + U^{n-1}f$, $f^0 = 0$, for all $f \in L^1_{\mathbb{R}}(X)$. If $E_N = \{x : \max_{0 \le n \le N} f^n > 0\}$ then* $\int_{E_N} f \, dm \ge 0$.

Proof Let $F = \max_{0 \le n \le N} f^n$. Note that $f + Uf^n = f^{n+1}$ and $U\left(\max_{0 \le n \le N} f^n\right) \ge \max_{0 \le n \le N} Uf^n$. Hence

$$f + UF \ge \max_{0 \le n \le N} Uf^n + f = \max_{1 \le n \le N+1} f^n \ge \max_{1 \le n \le N} f^n.$$

Moreover on E_N, $f + UF \ge F$. Therefore

$$\int_{E_N} f \, dm \ge \int_{E_N} (F - UF) \, dm \ge \int_X F \, dm - \int_X UF \, dm \ge 0,$$

since $F \ge 0$ and $UF \ge 0$ and since $F = 0$ on $X - E_N$. The proof is complete.

Corollary *Let T be a measure-preserving transformation of $(X, \mathscr{B}, m)$. For $f \in L^1(X)$, $m\{x : \overline{\lim} |(1/n)\sum_{k=0}^{n-1} f(T^k x)| > \lambda \|f\|\} \le m\{x : \sup_{n \ge 1} |(1/n)\sum_{k=0}^{n-1} f(T^k x)| > \lambda \|f\|\} \le (1/\lambda)(\lambda > 0)$.*

Proof We may take $\|f\| = 1$ and by increasing the set on the left-hand side replace f by $|f|$ so that it suffices to assume that f is positive. Let

$$E_N = \{x : \max_{1 \le n \le N} (f(x) + \cdots + f(T^{n-1}x)) > n\lambda\}$$
$$= \{x : \max_{0 \le n \le N} g^n(x) > 0\} \quad \text{where } g = f - \lambda.$$

Then

$$\int_{E_N} g \, dm \ge 0 \quad \text{i.e.} \quad \int_{E_N} f \, dm \ge \lambda m(E_N).$$

Hence

$$\int f \, dm \ge \lambda m(E) \quad \text{where} \quad E = \bigcup_{N=1}^{\infty} E_N.$$

This proves the corollary.

9. Theorem *Birkhoff's 'individual' ergodic theorem* [1]. *If T is a measure-preserving map of* $(X, \mathcal{B}, m)$ *then for each* $f \in L^1(X)$

$$\frac{1}{N} \sum_{n=0}^{N-1} f(T^n x) \to E(f|\mathcal{I})(x) \quad \text{a.e.}$$

In particular, the limiting function has the same integral as f and is this integral when T is ergodic.

Proof By subtracting the limiting function from the left-hand side it suffices to show that $(1/N) \sum_{n=0}^{N-1} f(T^n x) \to 0$ a.e. for each $f \in \bar{B}$ = ker $E(\ |\mathcal{I})$. Of course this is obvious for $f \in B_0 = \{fT - f : f \in L^\infty(X, \mathcal{B}, m)\}$ and $\bar{B}_0 = \bar{B}$. Let $E(f|\mathcal{I}) = 0$ and define

$$R(x, f) = \overline{\lim} \left| \frac{1}{N} \sum_{n=0}^{N-1} f(T^n x) \right|.$$

Evidently $R(x, f) = R(x, f - g)$ a.e. for all $g \in B_0$ and

$$m\{x : R(x, f) > \lambda \|f - g\|\} = m\{x : R(x, f - g) > \lambda \|f - g\|\} \le (1/\lambda).$$

Choose $\varepsilon > 0$ and $g \in B_0$ with $\|f - g\| < \varepsilon^2$ and $\lambda = 1/\varepsilon$. Then $m\{x : R(x, f) > \varepsilon\} \le \varepsilon$. Since $\varepsilon > 0$ is arbitrary, we see that $R(x, f) = 0$ a.e., and the proof is complete.

This proof of Birkhoff's theorem (due essentially to Garsia [2]) using the maximal and decomposition theorems, is not necessarily the simplest available (although the proof of the maximal theorem is surely by now perfected), but we have presented it, as it seems to us the clearest in terms of the ideas it involves and it mimics as closely as possible the considerably easier proof of von Neumann's theorem.

Wiener considered the problem of how ergodic averages $(1/N) \sum_{n=0}^{N-1} f \circ T^n$ converge to $E(f|\mathcal{I})$. When are such averages dominated by an integrable function?

The function $\log^+$ is defined on the non-negative reals by

$$\log^+ x = \log x \quad \text{for} \quad x \ge 1 \text{ (and zero elsewhere)}.$$

We shall need the following well-known fact – an exercise in integration theory and basic to probability theory.

If $f \ge 0$ and $F(t) = m\{x : f(x) > t\}$ then $\int g(f(x)) dm = -\int_0^\infty g(t) dF(t)$ when these integrals are well defined.

10. Theorem *Wiener's dominated ergodic theorem* [1].

If $f \in L^p (p > 1)$ *then* $f^* = \sup_{N \geq 1} |(1/N) \sum_{n=0}^{N-1} f \circ T^n| \in L^p(X)$; *If* $\int |f| \log^+ |f| \, dm < \infty$ *then* $f^* \in L^1(X)$.

Proof First it is worth remarking on the simple, but relevant fact that the class of functions f with $\int |f| \log^+ |f| \, dm < \infty$ is contained in $L^1(X)$ and for all $p > 1$ contains $L^p(X)$. This class is also linear.

We have seen $m\{x : f^*(x) > \lambda\} \leq (1/\lambda) \int f \, dm$, if, as we may suppose, $f \geq 0$. Let $h = f$ if $f > \frac{1}{2}\lambda$ and let $h = 0$ otherwise. Then $f \leq h + \frac{1}{2}\lambda$ and $F^*(\lambda) = m\{x : f^*(x) > \lambda\} \leq m\{x : h^*(x) > \frac{1}{2}\lambda\} \leq (2/\lambda) \int h \, dm = (2/\lambda) \int_{[f > (1/2)\lambda]} f \, dm$.

Let $p \geq 0$; then

$$\int f^{*1+p} \, dm = - \int_0^\infty t^{1+p} \, dF^*(t)$$

$$= - [t^{1+p} F^*(t)]_0^\infty + \int_0^\infty F^*(t)(1+p)t^p \, dt,$$

as long as, for example, the latter integral is finite, which is the case when $\int_a^\infty F^*(t)(1+p)t^p \, dt < \infty$, for some $0 \leq a < \infty$.

$$\int_a^\infty F^*(t)(1+p)t^p \, dt \leq \int_a^\infty \frac{2}{t} t^p (1+p) \left(\int_{[f > (1/2)t]} f \, dm \right) dt$$

$$= - \int_a^\infty 2t^{p-1}(1+p) \left(\int_{(1/2)t}^\infty s \, dF(s) \right) dt.$$

The region of double integration is

$$\{(s,t) : t \geq a, s \geq \tfrac{1}{2}t\} = \{(s,t) : s \geq \tfrac{1}{2}a, a \leq t \leq 2s\}$$

and Fubini's theorem applies since the integrand is non-negative. Hence

$$\int_a^\infty F^*(t)(1+p)t^p \, dt \leq - \int_{(1/2)a}^\infty s \left(\int_a^{2s} 2t^{p-1}(1+p) \, dt \right) dF(s)$$

$$= I_p = - \int_{(1/2)a}^\infty 2(1+p)s \left[\frac{t^p}{p} \right]_a^{2s} dF(s) \quad \text{when} \quad p > 0$$

$$\left(= I_0 = \int_{(1/2)a}^\infty 2s [\log t]_a^{2s} \, dF(s) \quad \text{when} \quad p = 0 \right).$$

For $p > 0$ we take $a = 0$ and conclude that

$$\int f^{*1+p}\,dm \leq I_p = -\int_0^\infty \frac{2(1+p)}{p}(2s)^p s\,dF(s)$$

$$= \frac{2(1+p)}{p}\left[-2^p \int_0^\infty s^{p+1}\,dF(s)\right]$$

$$= \frac{2(1+p)}{p}\left(2^p \int f^{p+1}\,dm\right) < \infty,$$

if $\int f^{p+1}\,dm < \infty$.

For $p = 0$ we take $a = 1$ and conclude that

$$\int f^*\,dm \leq I_0 = -\int_{(1/2)}^\infty 2s\log 2s\,dF(s)$$

$$= \int_{[f > 1/2]} 2f\log 2f\,dm$$

$$= \int 2f\log^+ 2f\,dm.$$

If $\int f\log^+ f\,dm < \infty$ then $\int 2f\log^+ 2f\,dm < \infty$, hence $\int f^*\,dm < \infty$ when $\int f\log^+ f\,dm < \infty$. The theorem is proved.

Exercise 16   Show that $f(T^n x)/n \to 0$ a.e. when $f \in L^1(X, \mathscr{B}, m)$. Seek (also) proofs independent of the maximal lemma and Birkhoff's theorem.

Exercise 17   A sequence of real integrable functions $\{f_n\}$ defined on $(X, \mathscr{B}, m)$ is said to *identically distributed* if $m(f_n^{-1}B)$ is independent of n, for all Borel subsets B of the real line and *independent* if $m(f_1^{-1}B_1 \cap \ldots \cap f_n^{-1}B_n) = m(f_1^{-1}B_1)\ldots m(f_n^{-1}B_n)$ for all n and all Borel subsets of $\mathbb{R}$. Note that $p(B_1 \times \ldots \times B_n) = m(f_1^{-1}B_1 \cap \ldots \cap f_n^{-1}B_n)$ generates a probability defined on $\prod_{n=1}^\infty \mathbb{R}$, and show that p is T-invariant where $T\{y_n\} = \{y_{n+1}\}$.

Exercise 18   Assuming that T (of Exercise 17) is ergodic (see Chapter 2. Exercise 4), prove the *strong law of large numbers*

$$\frac{1}{N}\sum_{n=1}^N f_n(x) \to \int f_1(x)\,dm \quad \text{(a.e.)}.$$

(Show that $x \to \varphi(x) = (f_1(x), f_2(x), \ldots)$ is measure-preserving, i.e. $m\varphi^{-1}(B) = p(B)$.)

Exercise 19 Let T be a measure-preserving transformation of the probability space $(X, \mathscr{B}, m)$. Prove the L^p mean ergodic theorem for $1 \leq p < \infty$. For $f \in L^p(X, \mathscr{B}, m)$ the following L^p convergence holds:

$$\frac{1}{N} \sum_{n=0}^{N-1} f \circ T^n \to E(f|\mathscr{I}) \in L^p(X, \mathscr{B}, m).$$

Exercise 20 For $1 < p < \infty$, deduce the L^p mean ergodic theorem (see Exercise 19) from Birkhoff's individual ergodic theorem and Wiener's dominated ergodic theorem.

Exercise 21 Let $\{T_t : t \in \mathbb{R}\}$ be a one-parameter group of measure-preserving transformations of $(X, \mathscr{B}, m)$ such that the map $\mathbb{R} \times X \to X$, $(t, x) \to T_t x$ is measurable. Show that L^p mean ergodic theorems (for $1 \leq p < \infty$) and individual ergodic theorems hold:

$$\frac{1}{T} \int_0^T f(T_t) \, dt \to E(f|\mathscr{I}).$$

Hint: Define $g(x) = \int_0^1 f(T_t x) \, dt$ and consider the transformation T_1 applied to g.

Exercise 22 Show that $\{f : \int |f| \log^+ |f| \, dm < \infty\}$ is a linear subspace of $L^1(X, \mathscr{B}, m)$ and contains each $L^p(X, \mathscr{B}, m)$ for $p > 1$.

2
Martingales and the ergodic theorem of information theory

1 Martingales

Let f be an integrable function defined on the unit interval. If we divide the unit interval into consecutive subintervals of length $1/2^n$ and replace f by the function f_n which has value $2^n \int_{i/2^n}^{(i+1)/2^n} f \, dx$ on the interval $[i/2^n, (i+1)/2^n)$ (in other words we give f_n the mean value of f over $[i/2^n, (i+1)/2^n)$ on this interval), we expect f_n to be a good approximation to f for large n. Writing $\mathscr{B}_n$ for the finite σ-algebra generated by the intervals $[i/2^n, (i+1)/2^n)$, f_n is nothing but $E(f/\mathscr{B}_n)$ and indeed it is true that $E(f|\mathscr{B}_n) \to f$ a.e. and in $L^1([0,1))$. This is an application of one of the Martingale theorems. (cf. Doob [1] for a general reference). The resemblance of the following maximal lemma to the maximal ergodic theorem will be noticed.

1. Maximal lemma *If* $(X, \mathscr{B}, m)$ *is a probability space and if* $\mathscr{B}_1 \subset \mathscr{B}_2 \subset \cdots \subset \mathscr{B}_N \subset \mathscr{B}$ *are* σ-algebras *then* *for* $\lambda > 0$, $E = \{x : \max_{1 \le n \le N} E(f|\mathscr{B}_n) > \lambda\}$, *we have, when* $f \in L^1(X)$,

$$m(E) \le (1/\lambda) \int |f| \, dm.$$

Proof We can take f to be non-negative. Let

$$E_n = \{x : E(f|\mathscr{B}_n) > \lambda, \quad E(f|\mathscr{B}_i) \le \lambda, \quad i = 1, 2, \ldots, n-1\}$$

so that $E = E_1 \cup \cdots \cup E_N$ where these sets are pairwise disjoint and $E_n \in \mathscr{B}_n$.

$$\int_E f \, dm = \sum_{n=1}^{N} \int_{E_n} f \, dm = \sum_{n=1}^{N} \int_{E_n} E(f|\mathscr{B}_n) \, dm$$

$$\ge \sum_{n=1}^{N} \lambda m(E_n) = \lambda m(E).$$

Hence

$$m(E) \le (1/\lambda) \int f \, dm.$$

2. Theorem *Increasing Martingale theorem. If* $\{\mathscr{B}_n\}$ *is an increasing sequence of sub-σ-algebras with* $\mathscr{B}_n \uparrow \mathscr{B}_\infty$ *(i.e.* $\bigcup_n \mathscr{B}_n$ *generates* $\mathscr{B}_\infty$*), then* $E(f|\mathscr{B}_n) \to E(f|\mathscr{B}_\infty)$ *a.e. and in* $L^1(X)$, *for* $f \in L^1(X)$.

Proof By replacing f by $E(f|\mathscr{B}_\infty)$ we may suppose $\mathscr{B}_\infty = \mathscr{B}$ and $E(f|\mathscr{B}_\infty) = f$. This theorem is true for a dense subspace of $L^1(X, \mathscr{B}, m)$, namely for functions in $\bigcup_n L^1(X, \mathscr{B}_n, m)$. This is because $f \in L^1(X, \mathscr{B}_k, m)$ implies $E(f|\mathscr{B}_n) = f$ for $n \geq k$.

Let $f \in L^1(X, \mathscr{B}, m)$ and choose $\varepsilon > 0$, $g \in L^1(X, \mathscr{B}_k, m)$ (some k) with $\int |f - g| \, dm < \varepsilon$. Then

$$\int |E(f|\mathscr{B}_n) - f| \, dm$$
$$\leq \int |E(f|\mathscr{B}_n) - E(g|\mathscr{B}_n)| \, dm + \int |E(g|\mathscr{B}_n) - g| \, dm + \int |g - f| \, dm$$
$$\leq 2 \int |g - f| \, dm \quad \text{if } n \geq k.$$

Hence

$$\overline{\lim} \int |E(f|\mathscr{B}_n) - f| \, dm \leq 2\varepsilon.$$

Since $\varepsilon > 0$ is arbitrary, $E(f|\mathscr{B}_n) \to f$ in $L^1(X)$. The maximal lemma establishes a.e. convergence as follows:

$$m\{x : \overline{\lim} |E(f|\mathscr{B}_n) - f| > \varepsilon^{1/2}\}$$
$$\leq m\{x : \overline{\lim} |E((f - g)|\mathscr{B}_n) - (f - g)| + |E(g|\mathscr{B}_n) - g| > \varepsilon^{1/2}\}$$
$$\leq m\{x : \overline{\lim} |E((f - g)|\mathscr{B})| + |f - g| > \varepsilon^{1/2}\}$$
$$\leq m\{x : \sup_n |E((f - g)|\mathscr{B}_n)| > \tfrac{1}{2}\varepsilon^{1/2}\} + m\{x : |f - g| > \tfrac{1}{2}\varepsilon^{1/2}\}$$
$$\leq (2/\varepsilon^{1/2}) \int |f - g| \, dm + (2/\varepsilon^{1/2}) \int |f - g| \, dm \leq 4\varepsilon^{1/2}.$$

Since $\varepsilon > 0$ is arbitrary,

$$\overline{\lim} |E(f|\mathscr{B}_n) - f| = 0 \quad \text{a.e.}$$

3. Theorem *Decreasing Martingale theorem. If* $\mathscr{B}_1 \supset \mathscr{B}_2 \supset \cdots$ *is a decreasing sequence of sub-σ-algebras such that* $\mathscr{B}_n \downarrow \mathscr{B}_\infty$ *(i.e.* $\bigcap_n \mathscr{B}_n = \mathscr{B}_\infty$*), then* $E(f|\mathscr{B}_n) \to E(f|\mathscr{B}_\infty)$ *a.e. and in* $L^1(X)$, *when* $f \in L^1(X)$.

Proof Since $\mathscr{B}_N \subset \mathscr{B}_{N-1} \subset \cdots \subset \mathscr{B}_1$ we can use the maximal inequality to get

$$m\{x : \sup_{1 \leq n \leq N} E(f|\mathscr{B}_n) > \lambda\} \leq (1/\lambda) \int |f| \, dm$$

for $f \in L^1(X)$ $(\lambda > 0)$, and this implies

$$m\{x: \sup_{1 \le n \le \infty} E(f|\mathcal{B}_n) > \lambda\} \le (1/\lambda) \int |f| \, dm$$

and $m\{x: \overline{\lim} E(f|\mathcal{B}_n) > \lambda\} \le (1/\lambda) \int |f| \, dm$.

Let $V_n = \ker E(\quad |\mathcal{B}_n) = \{f \in L^1(X) : E(f|\mathcal{B}_n) = 0\}$ then $V_1 \subset V_2 \cdots$. Let us show that $\bigcup_n V_n$ is dense in $V_\infty = \ker E(\quad |\mathcal{B}_\infty)$. To do this it will suffice to show that any continuous linear functional which annihilates $\bigcup_n V_n$ also annihilates V_∞. But such functionals have the form

$$J : f \to \int f h \, dm$$

where $h \in L^\infty(X, \mathcal{B}, m)$. Suppose J annihilates V_n for all n, then J annihilates all functions of the form $f - E(f|\mathcal{B}_n)$, i.e. $\int (f - E(f|\mathcal{B}_n)) h \, dm = 0$ for all n, when $f \in L^1(X)$. In particular,

$$\int (h - E(h|\mathcal{B}_n)) h \, dm = 0$$

and

$$\int (h - E(h|\mathcal{B}_n)) E(h|\mathcal{B}_n) \, dm = 0$$

is *always* true (for $h \in L^\infty(X)$). Subtracting, we have

$$\int (h - E(h|\mathcal{B}_n))^2 \, dm = 0$$

i.e. $h = E(h|\mathcal{B}_n) \in L^\infty(X, \mathcal{B}_n, m)$. Therefore $h \in L^\infty(X, \mathcal{B}_\infty, m)$. We now see that $E(f|\mathcal{B}_\infty) = 0$ implies

$$\int f h \, dm = \int E(f h|\mathcal{B}_\infty) \, dm = \int h E(f|\mathcal{B}_\infty) \, dm = 0$$

and we have proved that J annihilates V_∞ when J annihilates $\bigcup_n V_n$. In other words, $\bigcup_n V_n$ is dense in V_∞.

We note that the theorem is true for the dense subspace V of $L^1(X, \mathcal{B}, m)$ where $V = L^1(X, \mathcal{B}_\infty, m) + \bigcup_n V_n$.

Let $f \in L^1(X, \mathcal{B}, m)$ and choose $\varepsilon > 0$ and $g \in V$ with $\int |f - g| \, dm < \varepsilon$. Then

$$\int |E(f|\mathcal{B}_n) - E(f|\mathcal{B}_\infty)| \, dm$$
$$\le \int |E((f - g)|\mathcal{B}_n) - E((f - g)|\mathcal{B}_\infty)| \, dm$$
$$+ \int |E(g|\mathcal{B}_n) - E(g|\mathcal{B}_\infty)| \, dm$$
$$\le 2 \int |f - g| \, dm + \int |E(g|\mathcal{B}_n) - E(g|\mathcal{B}_\infty)| \, dm.$$

Hence

$$\overline{\lim} \int |E(f|\mathcal{B}_n) - E(f|\mathcal{B}_\infty)| \, dm \le 2 \int |f - g| \, dm \le 2\varepsilon.$$

Since $\varepsilon > 0$ is arbitrary, $L^1(X)$ convergence is established. To prove a.e. convergence we note that

$$m\{x : \overline{\lim} |E(f|\mathscr{B}_n) - E(f|\mathscr{B}_\infty)| > \varepsilon^{1/2}\}$$

$$\leq m\{x : \overline{\lim} |E((f-g)|\mathscr{B}_n) - E((f-g)|\mathscr{B}_\infty)|$$
$$+ \overline{\lim} |E(g|\mathscr{B}_n) - E(g|\mathscr{B}_\infty)| > \varepsilon^{1/2}\}$$

$$\leq m\{x : \sup_n |E((f-g)|\mathscr{B}_n) - E((f-g)|\mathscr{B}_\infty)| > \varepsilon^{1/2}\}$$

$$\leq m\{x : \sup_n |E((f-g)|\mathscr{B}_n)| > \tfrac{1}{2}\varepsilon^{1/2}\}$$

$$+ m\{x : \sup_n |E((f-g)|\mathscr{B}_\infty)| > \tfrac{1}{2}\varepsilon^{1/2}\}$$

$$\leq (2/\varepsilon^{1/2}) \int |f-g| \, dm + (2/\varepsilon^{1/2}) \int |f-g| \, dm \leq 4\varepsilon^{1/2}.$$

Since $\varepsilon > 0$ is arbitrary we conclude that

$$\overline{\lim} |E(f|\mathscr{B}_n) - E(f|\mathscr{B}_\infty)| = 0 \quad \text{a.e.}$$

Exercise 1 If $(X, \mathscr{B}, m)$ is a probability space and $\alpha = (A_1, A_2, \ldots)$ is a countable partition of X into measurable sets ($A_i \cap A_j = \varnothing$, $i \neq j$ and $\bigcup_{n=1}^\infty A_n = X$), show that

$$E(f|\mathscr{A})(x) = \frac{1}{m(A)} \int_A f \, dm \quad \text{for} \quad x \in A \in \alpha$$

when $f \in L^1(X, \mathscr{B}, m)$ and $\mathscr{A}$ is the σ-algebra generated by α.

Exercise 2 If $(X, \mathscr{B}, m), (X', \mathscr{B},' m')$ are two probability spaces show that

$$\int_{X'} f(x, y) \, dm'(y) = E(f|\mathscr{B} \times \mathscr{N}') \quad \text{a.e.}$$

when $f \in L^1(X \times X')$ and $\mathscr{N}'$ is the trivial sub-σ-algebra of $\mathscr{B}'$ consisting of sets of measure zero and one.

Exercise 3 Let $(X_\infty, \mathscr{B}_\infty, m_\infty) = \prod_{n=0}^\infty (X, \mathscr{B}, m)$ where $(X, \mathscr{B}, m)$ is a probability space. If T is the shift on X_∞, $T\{x_n\} = \{x_{n+1}\}$, then T is m_∞–preserving and $\mathscr{B}_\infty \supset T^{-1}\mathscr{B}_\infty \supset T^{-2}\mathscr{B}_\infty \supset \cdots$. Show that $\mathscr{B}' = \bigcap_n T^{-n}\mathscr{B}_\infty$ is trivial. (Take $f \in L^1(X_\infty, \mathscr{B}_\infty, m_\infty)$ dependent on only a finite number of coordinates and show that $E(f|T^{-n}\mathscr{B}_\infty) \to \int f \, dm$. Deduce the same result for *any* $f \in L^1(X_\infty, \mathscr{B}_\infty, m_\infty)$.)

Exercise 4 Show that the transformation T in Exercise 18 (Chapter 1) is ergodic by appealing to Exercise 3 above. Note that the strong law of large numbers (Exercise 18, Chapter 1) is now proved.

2 Information

Information is a function defined (on a probability space) with respect to a finite or countable partition. Its mean value, entropy, is thus a function of the partition. There are many axiomatic characterisations of entropy showing that it is necessarily defined as it is, given that it should obey certain more or less natural rules. (See, for example, Kendall [1]; Lee [1]; Rényi [1]; Aczél and Daróczy [1].) We choose to start with information and define entropy as its integral. Information can be motivated as follows: suppose we are trying to locate a point in a probability space $(X, \mathcal{B}, m)$ and to do this we use a 'grid' or partition $\alpha = (A_1, A_2, \ldots)$ of measurable subsets of X. On discovering that $x \in A_i$ (a specific element of α) we are in fact receiving information. Thus we view information (with respect to α) as a function which, when evaluated at x, depends only on the size of A_i, the element of α to which x belongs.

In other words

$$I(\alpha)(x) = \sum_{A \in \alpha} \chi_A(x) \varphi(m(A)).$$

If we use another 'independent' partition β it is natural to require that

$$I(\alpha \vee \beta) = I(\alpha) + I(\beta)$$

where $\alpha \vee \beta = \{A \cap B : A \in \alpha, B \in \beta\}$, i.e. the total information received using both partitions is the sum of the information received using each partition separately. Hence for two independent sets A, B $(m(A \cap B) = mA \cdot mB)$ we require that

$$\varphi(m(A \cap B)) = \varphi(m(A) \cdot m(B)) = \varphi(m(A)) + \varphi(m(B)).$$

If we also suppose that φ is continuous we are forced to the conclusion that $\varphi(t)$ is a constant multiple of $-\log t$. Hence we define

$$I(\alpha)(x) = - \sum_{A \in \alpha} \chi_A(x) \log m(A).$$

More generally, if ξ is a countable partition of the probability space $(X, \mathscr{B}, m)$, we define the *conditional information of* ξ *given* $\mathscr{A}$ (for a sub-σ-algebra $\mathscr{A}$) as

$$I(\xi|\mathscr{A}) = - \sum_{A \in \xi} \chi_A \log m(A/\mathscr{A}).$$

The *conditional entropy of* ξ *given* $\mathscr{A}$ is

$$H(\xi|\mathscr{A}) = \int I(\xi|\mathscr{A})\,dm = \int - \sum_{A \in \xi} m(A|\mathscr{A}) \log m(A|\mathscr{A})\,dm.$$

The above information and entropy are unconditional when $\mathscr{A} = \mathscr{N}$, the trivial σ-algebra consisting of sets of measure zero and one, and in this case we write $I(\xi|\mathscr{N}) = I(\xi)$ (the *information* of ξ), and $H(\xi|\mathscr{N}) = H(\xi)$ (the *entropy* of ξ). As is easily seen, $H(\xi|\mathscr{A}) = 0$ (*or* $I(\xi|\mathscr{A}) \equiv 0$) *if and only if* ξ *consists of* $\mathscr{A}$ *sets*. In particular, $H(\xi|\mathscr{B}) = 0$ and $I(\xi|\mathscr{B}) = 0$ for any countable (and of course measurable) partition ξ. At this point it is convenient to introduce the following convention: *if* ξ *is a partition then* $\hat{\xi}$ *denotes the* σ*-algebra generated by* ξ.

For countable partitions ξ, α we shall sometimes write $I(\xi|\alpha)$ and $H(\xi|\alpha)$ for $I(\xi|\hat{\alpha})$ and $H(\xi|\hat{\alpha})$, respectively.

If $\xi_1, \ldots, \xi_n$ are countable partitions, then $\xi_1 \vee \cdots \vee \xi_n = \bigvee_{i=1}^{n} \xi_i$ denotes the partition consisting of sets $A_1 \cap \cdots \cap A_n$ with $A_i \in \xi_i$. Similarly for σ-algebras $\mathscr{A}_1, \mathscr{A}_2, \ldots, \mathscr{A}_1 \vee \mathscr{A}_2 \vee \cdots$ or $\bigvee_i \mathscr{A}_i$ denotes the σ-algebra generated by $\bigcup_n \mathscr{A}_n$. Evidently $\widehat{\bigvee_i \xi_i} = \bigvee_i \hat{\xi}_i$ for partitions $\xi_1, \ldots, \xi_n$;

If ξ, η are two partitions, then $\xi \geq \eta$ means every set in η is a union of sets in ξ or, equivalently, $\xi = \xi \vee \eta$.

For three countable partitions ξ, η, ζ we have the following:

Basic identities

$$I(\xi \vee \eta|\zeta) = I(\xi|\zeta) + I(\eta|\xi \vee \zeta)$$

and

$$H(\xi \vee \eta|\zeta) = H(\xi|\zeta) + H(\eta|\xi \vee \zeta).$$

The second of these identities is obtained from the first by integration. To prove the first note that $I(\xi \vee \eta|\zeta)(x) = -\log m(A \cap B|\zeta)$ for

$x \in A \cap B$, $A \in \xi$, $B \in \eta$ and

$$m(A \cap B | \zeta) = \sum_{C \in \zeta} \chi_C \frac{m(A \cap B \cap C)}{m(C)}$$

(see Exercise 1 of this Chapter). Thus

$$I(\xi \vee \eta | \zeta)(x) = -\log \frac{m(A \cap B \cap C)}{m(C)}$$

for $x \in A \cap B \cap C$, $A \in \xi$, $B \in \eta$, $C \in \zeta$. On the other hand,

$$I(\xi | \zeta)(x) = -\log \frac{m(A \cap C)}{m(C)}$$

for $x \in A \cap C$, $A \in \xi$, $C \in \zeta$ and

$$I(\eta | \xi \vee \zeta)(x) = -\log \frac{m(B \cap A \cap C)}{m(A \cap C)}$$

for $x \in A \cap B \cap C$, $A \in \xi$, $B \in \eta$, $C \in \zeta$. Hence

$$I(\xi | \zeta)(x) + I(\eta | \xi \vee \zeta)(x) = -\log \frac{m(A \cap B \cap C)}{m(C)} = -I(\xi \vee \eta | \zeta)(x)$$

for $x \in A \cap B \cap C$, and the proof is complete.

As consequences of the above we have

$$I(\xi \vee \eta | \zeta) = I(\xi | \zeta) \quad \text{and} \quad H(\xi \vee \eta | \zeta) = H(\xi | \zeta) \quad \text{when} \quad \eta \leq \zeta.$$

From the basic identities we deduce that

$$I(\xi | \zeta) \geq I(\eta | \zeta) \quad \text{and} \quad H(\xi | \zeta) \geq H(\eta | \zeta)$$

when $\xi \geq \eta$. Not so immediate is the inequality

$$H(\xi | \alpha) \geq H(\xi | \beta) \quad \text{when} \quad \beta \geq \alpha;$$

the proof depends on Jensen's inequality. However, the inequality $I(\xi | \alpha) \geq I(\xi | \beta)$ (when $\beta \geq \alpha$) is *not generally true*.

4. Theorem *Jensen's inequality. Let φ be a continuous concave function defined on the unit interval $[0, 1]$ i.e. $\varphi(sx + ty) \geq s\varphi(x) + t\varphi(y)$ whenever $0 \leq s, t \leq 1$, $s + t = 1$. Let $f : (X, \mathcal{B}, m) \to [0, 1]$, then for any sub-$\sigma$-algebra $\mathcal{A}$ of $\mathcal{B}$ we have*

$$\varphi(E(f | \mathcal{A})) \geq E(\varphi(f) | \mathcal{A}) \quad \text{a.e.}$$

Proof Suppose f is a simple function mapping X into $[0, 1]$; $f(x) = \sum_i b_i \chi_{B_i}$ where $\{B_i\}$ is a partition of X. Then

$$\sum_i m(B_i | \mathscr{A}) = 1 \quad \text{and} \quad \sum_{i\cdot} m(B_i | \mathscr{A}) \varphi(b_i) \le \varphi \left(\sum_i b_i m(B_i | \mathscr{A}) \right)$$

(concavity), i.e. $E(\varphi(f) | \mathscr{A}) \le \varphi(E(f | \mathscr{A}))$, for simple functions f.

Since each measurable f mapping X into $[0, 1]$ is the limit of an increasing sequence of simple functions f_n mapping X into $[0, 1]$, we deduce the general result using dominated convergence and continuity of φ.

Corollary $H(\xi | \mathscr{A}_1) \le H(\xi | \mathscr{A}_2)$ *when ξ is a countable partition and $\mathscr{A}_1$, $\mathscr{A}_2$ are sub-σ-algebras with $\mathscr{A}_1 \supset \mathscr{A}_2$.*

Proof Let $\varphi(t) = -t \log t$ $(t \neq 0)$, $\varphi(0) = 0$. φ is a continuous concave function on $[0, 1]$. Let $f(x) = m(F | \mathscr{A}_1)(x)$ $(F \in \xi)$, then

$$\varphi(E(f | \mathscr{A}_2)) \ge E(\varphi(f) | \mathscr{A}_2),$$
$$\varphi(m(F | \mathscr{A}_2)) \ge E(-m(F | \mathscr{A}_1) \log m(F | \mathscr{A}_1) | \mathscr{A}_2).$$

Integrating we obtain

$$\int -m(F | \mathscr{A}_2) \log m(F | \mathscr{A}_2) \, dm \ge \int -m(F | \mathscr{A}_1) \log m(F | \mathscr{A}_1) \, dm$$

and summing over $F \in \xi$ we have

$$H(\xi | \mathscr{A}_2) \ge H(\xi | \mathscr{A}_1).$$

Exercise 5 Two countable partitions α, β are said to be *independent* if $m(A \cap B) = m(A) \cdot m(B)$ for all $A \in \alpha$, $B \in \beta$. Show that if $H(\alpha) < \infty$, $H(\beta) < \infty$, then α, β are independent if and only if $H(\alpha \vee \beta) = H(\alpha) + H(\beta)$, i.e. $H(\alpha | \beta) = H(\alpha)$. (Consider the inequalities

$$\sum_{B \in \beta} -m(B) \frac{m(A \cap B)}{m(B)} \log \frac{m(A \cap B)}{m(B)} \le -m(A) \log m(A)$$

for each $A \in \alpha$, and use the 'strict concavity' of φ above.)

Exercise 6 If α, β, γ are three countable partitions with $H(\alpha) < \infty$, $H(\beta) < \infty$, $H(\gamma) < \infty$, show that

$$H(\alpha | \beta \vee \gamma) = H(\alpha | \gamma) \quad \text{if and only if} \quad \frac{m(A \cap B \cap C)}{m(B \cap C)} = \frac{m(A \cap C)}{m(C)}$$

for all $A \in \alpha$, $B \in \beta$, $C \in \gamma$. (Apply Exercise 5 above to each probability space (C, m_C) where $m_C(E) = (m(E \cap C)/m(C))$.)

We shall need the following:

5. Lemma (Chung [1]) *If ξ is a countable partition with $H(\xi) < \infty$ and if $\mathcal{B}_1 \subset \mathcal{B}_2 \subset \cdots$ is an increasing sequence of sub-σ-algebras then*

$$\int \sup_n I(\xi | \mathcal{B}_n) \, dm \leq H(\xi) + 1.$$

Proof The proof resembles the proof of Wiener's dominated ergodic theorem.

Let $f = \sup_n I(\xi | \mathcal{B}_n)$ and define $F(t) = m\{x : f(x) > t\}$ so that

$$\int f \, dm = - \int_0^\infty t \, dF(t)$$

$$= [-tF(t)]_0^\infty + \int_0^\infty F(t) \, dt \leq \int_0^\infty F(t) \, dt.$$

$$F(t) = m\{x : \sup_n - \sum_{A \in \xi} \chi_A \log m(A | \mathcal{B}_n) > t\}$$

$$= \sum_{A \in \xi} m(A \cap \{x : \inf_n m(A | \mathcal{B}_n) < e^{-t}\}).$$

But $m(A \cap \{x : \inf_n m(A | \mathcal{B}_n) < e^{-t}\})$

$$= \sum_{n=1}^\infty m(A \cap \{x : m(A | \mathcal{B}_n) < e^{-t}, \, m(A | \mathcal{B}_k) \geq e^{-t}, \, k < n\})$$

$$= \sum_{n=1}^\infty \int_{A_n} m(A | \mathcal{B}_n) \, dm$$

where $A_n = \{x : m(A | \mathcal{B}_n) < e^{-t}, \, m(A | \mathcal{B}_k) \geq e^{-t}, \, k < n\}$. Since $\int_{A_n} m(A | \mathcal{B}_n) \, dm \leq e^{-t} m(A_n)$, we have

$$\sum_{n=1}^\infty m(A \cap A_n) \leq e^{-t}.$$

Hence

$$F(t) \leq \sum_{A \in \xi} \min(m(A), e^{-t})$$

and

$$\int_0^\infty F(t)\,dt \le \sum_{A \in \xi} \int_0^\infty \min{(m(A), e^{-t})}dt$$

$$= \sum_{A \in \xi} (-m(A)\log m(A) + m(A))$$

$$= H(\xi) + 1.$$

We conclude that $\int f\,dm \le H(\xi) + 1$.

6. Theorem *If ξ is a countable partition with $H(\xi) < \infty$ and if $\mathscr{B}_1 \subset \mathscr{B}_2 \subset \cdots$ is an increasing sequence of sub-σ-algebras with $\mathscr{B}_n \uparrow \mathscr{B}_\infty$ (i.e. $\mathscr{B}_\infty$ is generated by $\bigcup_n \mathscr{B}_n$), then $I(\xi|\mathscr{B}_n) \to I(\xi|\mathscr{B}_\infty)$ a.e. and in $L^1(X)$. Moreover, $H(\xi|\mathscr{B}_n) \downarrow H(\xi|\mathscr{B}_\infty)$.*

Proof For each $A \in \xi$, $m(A|\mathscr{B}_n) \to m(A|\mathscr{B}_\infty)$ a.e. by the Increasing Martingale theorem, hence $I(\xi|\mathscr{B}_n) \to I(\xi|\mathscr{B}_\infty)$ a.e., and since $\{I(\xi|\mathscr{B}_n)\}$ is dominated by an integrable function, by the preceding lemma, we have $I(\xi|\mathscr{B}_n) \to I(\xi|\mathscr{B}_\infty)$ in $L^1(X)$ also. Integration yields $H(\xi|\mathscr{B}_n) \to H(\xi|\mathscr{B}_\infty)$ and we know that the sequence $\{H(\xi|\mathscr{B}_n)\}$ is decreasing.

Corollary *If $(X, \mathscr{B}, m)$ is separable, then $I(\xi \vee \eta|\mathscr{A}) = I(\xi|\mathscr{A}) + I(\eta|\overset{\sim}{\xi} \vee \mathscr{A})$ and $H(\xi \vee \eta|\mathscr{A}) = H(\xi|\mathscr{A}) + H(\eta|\overset{\sim}{\xi} \vee \mathscr{A})$ for countable partitions ξ, η (with $H(\xi) < \infty$, $H(\eta) < \infty$) and sub-σ-algebras $\mathscr{A}$.*

Proof There exists a sequence of finite partitions $\{\alpha_n\}$ such that $\hat{\alpha}_n \uparrow \mathscr{A}$ (see Exercise 10 of this Chapter) so that taking limits of

$$I(\xi \vee \eta|\alpha_n) = I(\xi|\alpha_n) + I(\eta|\xi \vee \alpha_n)$$

the corollary is proved.

So far this chapter has diverged from any considerations of measure-preserving transformations and it is now time to re-introduce them. We first note the effect of a measure-preserving transformation T of $(X, \mathscr{B}, m)$ (onto itself) on the information function $I(\xi|\mathscr{A})$, where ξ is a countable partition and $\mathscr{A}$ is a sub-σ-algebra. In fact,

$$I(\xi|\mathscr{A}) \circ T = I(T^{-1}\xi|T^{-1}\mathscr{A})$$

as is easily checked using Chapter 1 (property (vi)), and integration of this identity yields

$$H(\xi|\mathscr{A}) = H(T^{-1}\xi|T^{-1}\mathscr{A}).$$

If ξ is a countable partition with $H(\xi) < \infty$, we define the entropy of T with respect to ξ as

$$h(T, \xi) = H\left(\xi \Big| \bigvee_{n=1}^{\infty} T^{-n}\xi\right).$$

This quantity will be exploited in Chapter 4. For the moment we note its significance in the ergodic theorem of information theory which provides an estimate for the size of sets $A_{x_0} \cap T^{-1}A_{x_1} \cap \cdots \cap T^{-n}A_{x_n}, A_{xi} \in \xi$. When T is ergodic the size of these sets (in general) is roughly $\exp[-n\,h(T, \xi)]$. The theorem was proved at various levels of generality by Shannon [1]; McMillan [1]; Breiman [1]; Ionescu Tulcea [1]; Chung [1].

7. Theorem *Ergodic theorem of information theory. Let ξ be a countable partition of the probability space $(X, \mathscr{B}, m)$ with $H(\xi) < \infty$ and let T be a measure-preserving transformation of $(X, \mathscr{B}, m)$. Writing $f = I(\xi|\bigvee_{n=1}^{\infty} T^{-n}\xi)$, we have*

$$\frac{1}{N}I\left(\bigvee_{n=0}^{N-1} T^{-n}\xi\right) \to E(f|\mathscr{I}) \quad a.e. \text{ and in } L^1(X)$$

and

$$\frac{1}{N}H\left(\bigvee_{n=0}^{N-1} T^{-n}\xi\right) \to h(T, \xi),$$

where $\mathscr{I}$ is the σ-algebra of T-invariant sets. When T is ergodic

$$\frac{1}{N}I\left(\bigvee_{n=0}^{N-1} T^{-n}\xi\right) \to H\left(\xi \Big| \bigvee_{n=1}^{\infty} T^{-n}\xi\right) = h(T, \xi) \quad a.e. \text{ and in } L^1(X).$$

Proof We note that $f = I(\xi|\bigvee_{n=1}^{\infty} T^{-n}\xi) \in L^1(X)$ and remark that it suffices to prove

$$\frac{1}{N}I\left(\bigvee_{n=0}^{N-1} \overline{T^{-n}\xi}\right) \to E(f|\mathscr{I}) \quad a.e. \text{ and in } L^1(X),$$

since integration yields the rest of the theorem.

Evidently,

$$I\left(\bigvee_{n=0}^{N-1} T^{-n}\xi\right) = I\left(\xi \bigg| \bigvee_{n=1}^{N-1} T^{-n}\xi\right) + I\left(\bigvee_{n=0}^{N-2} T^{-n}\xi\right) \circ T$$

$$= f_{N-1} + f_{N-2} \circ T + I\left(\bigvee_{n=0}^{N-3} T^{-n}\xi\right) \circ T^2$$

$$= f_{N-1} + f_{N-2} \circ T + \cdots + f_0 \circ T^{N-1}$$

where $f_m = I(\xi | \bigvee_{n=1}^m T^{-n}\xi)$, $f_0 = I(\xi)$. Moreover

$$\left|\frac{1}{N} I\left(\bigvee_{n=0}^{N-1} T^{-n}\xi\right) - E(f | \mathscr{I})\right| \le \frac{1}{N} \sum_{n=0}^{N-1} |f_{N-1-n} \circ T^n - f \circ T^n|$$

$$+ \left|\frac{1}{N} \sum_{n=0}^{N-1} f \circ T^n - E(f|\mathscr{I})\right|.$$

Since the last part of this expression converges to zero a.e. and in $L^1(X)$ (cf. Exercise 19, Chapter 1), we must prove

$$\overline{\lim}\frac{1}{N} \sum_{n=0}^{N-1} g_{N-1-n} \circ T^n = 0 \quad \text{a.e.}$$

and

$$\overline{\lim}\frac{1}{N} \sum_{n=0}^{N-1} \int g_{N-1-n} T^n \mathrm{d}m = 0 \quad \text{where } g_k = |f - f_k|.$$

But

$$\frac{1}{N} \sum_{n=0}^{N-1} \int g_{N-1-n} \circ T^n \, \mathrm{d}m = \frac{1}{N} \sum_{n=0}^{N-1} \int g_{N-1-n} \, \mathrm{d}m$$

$$= \frac{1}{N} \sum_{n=0}^{N-1} \int g_n \, \mathrm{d}m \to 0,$$

since

$$\int g_N \mathrm{d}m = \int \left|I\left(\xi \bigg| \bigvee_{n=1}^N T^{-n}\xi\right) - I\left(\xi \bigg| \bigvee_{n=1}^\infty T^{-n}\xi\right)\right| \mathrm{d}m \to 0$$

by Theorem 6. We proceed to the proof of a.e. convergence. Let

$$G_N = \sup_{n \ge N} g_n \le \sup_m I\left(\xi \bigg| \bigvee_{n=1}^m T^{-n}\xi\right) + I\left(\xi \bigg| \bigvee_{n=1}^\infty T^{-n}\xi\right) \in L^1(X)$$

by Lemma 5. Then $G_n \downarrow 0$ a.e., again by Theorem 6. Obviously,

$$\frac{1}{N} \sum_{n=0}^{N-1} g_{N-1-n} \circ T^n = \frac{1}{N} \sum_{n=0}^{N-M-1} g_{N-n-1} \circ T^n$$
$$+ \frac{1}{N} \sum_{n=N-M}^{N-1} g_{N-n-1} \circ T^n$$
$$\leq \frac{1}{N} \sum_{n=0}^{N-M-1} G_M \circ T^n + \frac{1}{N} \sum_{n=N-M}^{N-1} G_0 \circ T^n.$$

Taking limits and using Birkhoff's ergodic theorem we have

$$\overline{\lim} \frac{1}{N} \sum_{n=0}^{N-1} g_{N-1-n} \circ T^n \leq E(G_M | \mathcal{I}) \quad \text{a.e.,}$$

and this holds for all positive integers M. $E(G_M | \mathcal{I})$ decreases with M and

$$\int E(G_M | \mathcal{I}) \, dm = \int G_M \, dm \to 0,$$

therefore

$$\overline{\lim} \frac{1}{N} \sum_{n=0}^{N-1} g_{N-1-n} \circ T^n = 0 \quad \text{a.e.}$$

and the theorem is proved.

Exercise 7 Let T be an ergodic measure-preserving transformation of the probability space $(X, \mathcal{B}, m)$ and let ξ be a countable partition with $H(\xi) < \infty$. A ξ-*cylinder* of length $n+1$ is a set of the form $A_{x_0} \cap T^{-1} A_{x_1} \cap \cdots \cap T^{-n} A_{x_n}$, $A_{x_i} \in \xi$. Show that the ξ-cylinders C of length $n+1$ satisfying $\exp\{-n[h(T, \xi) - \varepsilon]\} \geq m(C) \geq \exp\{-n[h(T, \xi) + \varepsilon]\}$ exhaust all of X except for a set of measure at most ε, if n is large enough.

Exercise 8 Compute $h(T, \xi)$ for $Tx = nx \pmod 1$ (n a positive integer), $\xi = [0, 1/n), \ldots, [(n-1)/n, 1)$. ($T$ is Lebesgue measure-preserving.)

Exercise 9 Let $(X_\infty, \mathcal{B}_\infty, m_\infty) = \prod_{n=0}^\infty (X_n, \mathcal{B}_n, m_n)$ where $X_n = \{1, 2, \ldots\}$, $\mathcal{B}_n$ consists of all subsets of X_n and $m_n\{i\} = p_i$, $\sum_{i=1}^\infty p_i = 1$. If T is the shift transformation, $T\{x_n\} = \{x_{n+1}\}$ and $\xi = (A_1, A_2, \ldots)$, $A_i = \{x : x_0 = i\}$ show that $h(T, \xi) = -\sum_{i=1}^\infty p_i \log p_i$.

***Exercise* 10**   If $(X, \mathcal{B}, m)$ is a separable probability space and $\mathcal{A}$ is a sub-σ-algebra of $\mathcal{B}$, show that there exists a sequence $\{\alpha_n\}$ of finite partitions with $\hat{\alpha}_n \uparrow \mathcal{A}$. (By definition $(\mathcal{B}, d)$ is a separable metric space with $d(A, B) = m(A \triangle B)$. Therefore $(\mathcal{A}, d)$ is separable. Choose $\{A_n\}$, a countable dense sequence in $\mathcal{A}$, and construct the sequence $\{\alpha_n\}$ from $\{A_n\}$.)

***Exercise* 11**   If $(X, \mathcal{B}, m)$ is separable and $\mathcal{B}$ is the σ-algebra generated by the algebra $\mathcal{A}$, show that there is a sequence $\{\alpha_n\}$ of finite partitions with $\alpha_n \subset \mathcal{A}$ such that $\hat{\alpha}_n \uparrow \mathcal{B}$.

***Exercise* 12**   If $(X, \mathcal{B}, m)$ is separable and if $H(\xi) < \infty$, and $\{\mathcal{B}_n\}$ is a sequence of sub-σ-algebras such that $\mathcal{B}_n \uparrow \mathcal{B}$, show that for every $\varepsilon > 0$ there exists, for some n, $\eta \subset \mathcal{B}_n$ with $H(\eta) < \infty$ and $H(\xi|\eta) + H(\eta|\xi) < \varepsilon$. (Quite generally, show that $D(\alpha, \beta) = H(\alpha|\beta) + H(\beta|\alpha)$ is a metric on $Z = \{\alpha : H(\alpha) < \infty\}$. ξ may be assumed to be finite and is therefore the refinement of a finite sequence of two-set partitions. Prove Exercise 12 for two-set partitions.)

3
Mixing

1 Ergodicity

In this chapter we shall be using a number of facts from the spectral theory of unitary operators. We refer the reader to the Appendix for details.

If T is an ergodic measure-preserving transformation of the probability space $(X, \mathscr{B}, m)$, then by the ergodic theorem of von Neumann

$$\frac{1}{N} \sum_{n=0}^{N-1} \chi_A \circ T^n \to m(A) \tag{3.1}$$

for all $A \in \mathscr{B}$ and

$$\frac{1}{N} \sum_{n=0}^{N-1} f \circ T^n \to \int f \tag{3.2}$$

for all $f \in L^2(X, \mathscr{B}, m)$, where in each case the convergence is in the L^2 norm.

Taking inner products of (3.1) with χ_B and (3.2) with g we have

$$\frac{1}{N} \sum_{n=0}^{N-1} m(T^{-n} A \cap B) \to m(A) \cdot m(B) \tag{3.3}$$

for all $A, B \in \mathscr{B}$ and

$$\frac{1}{N} \sum_{n=0}^{N-1} \int (f \circ T^n) \bar{g} \, dm = \frac{1}{N} \sum_{n=0}^{N-1} \langle U_T^n f, g \rangle \to \int f \, dm \int \bar{g} \, dm$$
$$= \langle f, 1 \rangle \langle 1, g \rangle \tag{3.4}$$

for all $f, g \in L^2(X, \mathscr{B}, m)$.

Of course (3.1) is a special case of (3.2), and (3.3) is a special case of (3.4). However, it is not difficult to see that (3.4) can be deduced from (3.3). One first shows that (3.3) implies (3.4) for simple functions f, g and an approximation argument does the rest. Hence ergodicity implies the equivalent properties (3.3) and (3.4). Conversely, (3.3) or

(3.4) implies that T is ergodic. For if $T^{-1}A = A$ then taking $B = A^c$ we have, from (3.3),

$$\frac{1}{N}\sum_{n=0}^{N-1} m(T^{-n}A \cap B) = 0 = m(A)\,m(B)$$

i.e. $m(A) = 0$ or 1.

1. Theorem *If T is a measure-preserving transformation, then T is ergodic if and only if* $(1/N)\sum_{n=0}^{N-1}\int (f \circ T^n)\,\bar{f}\,dm \to 0$ *for all $f \in L^2(X, \mathcal{B}, m)$ such that $f \perp \mathbb{C}$(i.e. $\int f\,dm = 0$).*

Proof If T is ergodic then the convergence above follows from (3.4). We wish to prove the converse; but first some remarks are in order. We write $L^2(X, \mathcal{B}, m) = H \oplus H^\perp$ where $H = \bigcap_{n \geq 0} U_T^n L^2(X, \mathcal{B}, m)$ and note that $H^\perp = \sum_{n=0}^\infty \bigoplus U_T^n V$ where

$$V = L^2(X, \mathcal{B}, m) \ominus U_T L^2(X, T^{-1}\mathcal{B}, m)$$
$$= \{f \in L^2(X, \mathcal{B}, m) : f \perp L^2(X, T^{-1}\mathcal{B}, m)\}.$$

Since the spaces $U_T^n V$, $n = 0, 1, \ldots$, are mutually orthogonal, it is clear that for $f \in U_T^i V$, $g \in U_T^j V$,

$$\frac{1}{N}\sum_{n=0}^{N-1} \langle U_T^n f, g \rangle \to 0.$$

By taking linear combinations of such f, g and using an approximation argument we have $(1/N)\sum_{n=0}^{N-1} \langle U_T^n f, g \rangle \to 0$ for all $f, g \in H^\perp$. Of course for $f \in H$, $g \in H^\perp$ (or *vice versa*) $(1/N)\sum_{n=0}^{N-1} \langle U_T^n f, g \rangle = 0$. We see then that in order to prove the theorem it suffices to show that

$$\frac{1}{N}\sum_{n=0}^{N-1} \langle U_T^n f, g \rangle \to 0 \quad \text{when} \quad f, g \in H \quad \text{and} \quad \int f\,dm = 0.$$

In fact it is only at this stage of the proof that the hypothesis is relevant. We note first that $U_T H = H$ and therefore U_T (restricted to H) is a unitary operator, i.e. U_T has an inverse. Let $f \in H$, $f \perp \mathbb{C}$, and define $Z(f)$ to be the closure of the linear span $S(f)$ of $\{U_T^n f : n = 0, \pm 1, \ldots\}$. If $g \in S(f)$ then the hypothesis of the theorem ensures that $(1/N)\sum_{n=0}^{N-1} \langle U_T^n f, g \rangle \to 0$. Hence the same is true for $g \in Z(f)$. On the other hand, if $g \perp Z(f)$ then $(1/N)\sum_{n=0}^{N-1} \langle U_T^n f, g \rangle = 0$. We conclude that $(1/N)\sum_{n=0}^{N-1} \langle U_T^n f, g \rangle \to 0$ for all $f \in H$ with $\int f\,dm = 0$ and all $g \in H$. The proof is complete.

At this point it is convenient to introduce:

Herglotz's Theorem (cf. Appendix) *If U is an isometry on the Hilbert space H and $x \in H$ then there is a unique finite Borel measure $\tilde{x}$ on the circle K such that $\langle U^n x, x \rangle = \int_K z^n \, d\tilde{x}$ for all $n = 0, 1, \ldots$.*

(If U is unitary then this relation holds for $n = -1, -2, \ldots$ also.) $\tilde{x}$ is called the *spectral measure* of x. Theorem 1 characterises ergodicity in terms of the spectral measures of functions orthogonal to the constants since ergodicity is equivalent to

$$\frac{1}{N} \sum_{n=0}^{N-1} \int_K z^n \, d\tilde{f} \to 0 \quad \text{for all } f \perp \mathbb{C}.$$

We have already noticed that the ergodicity of T is equivalent to U_T having 1 as a simple eigenvalue. By virtue of this fact, ergodicity is a 'spectral' property as is demonstrated equally by Theorem 1. To be precise on this point, a property of a measure-preserving transformation T is said to be *spectral* if whenever $U_T = W U_S W^{-1}$, for an invertible isometry W and a measure-preserving transformation S, then S also shares this property. The relation $U_T = W U_S W^{-1}$ says that U_T, U_S are *spectrally* or *unitarily equivalent*.

The *point* or *discrete spectrum* of a measure-preserving transformation is the set $H(T)$ of *eigenvalues* of U_T:

$$H(T) = \{\alpha \in \mathbb{C} : f \circ T = \alpha f \text{ for some } f, 0 \neq f \in L^2(X)\}.$$

Since $f \circ T = \alpha f$ implies that $\int |f \circ T|^2 \, dm = \int |f|^2 \, dm = |\alpha|^2 \int |f|^2 \, dm$, we see that $|\alpha| = 1$ when $\alpha \in H(T)$. In the same way we see that *eigenfunctions* corresponding to distinct eigenvalues are orthogonal. Hence, for a separable probability space $H(T)$ is a countable set, for $L^2(X)$ is separable.

If T is ergodic then $H(T)$ is a subgroup of K. For if $f \circ T = \alpha f, g \circ T = \beta g$ then $|f|, |g|$ are invariant and therefore constant. Hence $1/f$ and fg belong to $L^2(X)$ and α^{-1}, $\alpha\beta$ are eigenvalues. We note also that for T ergodic, eigenfunctions have constant absolute value.

If T is a measure-preserving transformation of the probability space $(X, \mathcal{B}, m)$, we define V_d to be the closure of the linear span of the set of all eigenfunctions of U_T. We shall see later (Appendix) that $V_d = \{f \in L^2(X) : \tilde{f} \text{ is purely atomic}\}$, and $V_c = V_d^\perp = \{f \in L^2(X) : \tilde{f} \text{ is}$

non-atomic}. Hence we can always decompose $L^2(X)$ in accordance with the *discrete* and *continuous spectra* of $U_T : L^2(X) = V_d \oplus V_c$. T or U_T is said to have *discrete spectrum* if $V_d = L^2(X)$. T or U_T is said to have *continuous spectrum* in $\mathbb{C}^\perp$ if $V_d = \mathbb{C}$ (i.e. the only eigenfunctions are constant). *Mixed spectrum* in $\mathbb{C}^\perp$ describes all other cases.

We have already met transformations with discrete spectrum, namely the translations of the k-dimensional torus given by $T(x_1, \ldots, x_k) = (x_1 + \alpha_1, \ldots, x_k + \alpha_k) \pmod 1$. In fact for each lattice point $h = (h_1, \ldots, h_k) \in \mathbb{Z}^k$ the exponential function $f_h(x) = \exp 2\pi i \langle h, x \rangle$ is an eigenfunction with eigenvalue $\exp 2\pi i \langle h, \alpha \rangle$, and clearly the linear span of exponential functions is dense in the Hilbert space of square integrable functions.

The transformation $T(x, y) = (x + \alpha, x + y) \pmod 1$ of the 2-torus provides us with an example of mixed spectrum. For V_d is spanned by the eigenfunctions $f_m(x, y) = \exp(2\pi i m x)$ and V_c is spanned by the functions $f_{m,k}(k \neq 0)$ where $f_{m,k}(x, y) = \exp[2\pi i(mx + ky)]$. Moreover, $\bar{f}_{m,k}$ is Lebesgue measure (and therefore continuous) since

$$\int\!\!\int f_{m,k} T^n(x, y) f_{m,k}(x, y) \, dx \, dy = \int\!\!\int \exp\{2\pi i[m(x + n\alpha) + k(\tfrac{n}{2}(n-1)\alpha + nx + y)]\} \cdot \exp[-2\pi i(mx + ky)] \, dx \, dy$$

which equals one when $n = 0$ and zero otherwise. In other words,

$$\int\!\!\int f_{m,k} T^n(x, y) \bar{f}_{m,k}(x, y) \, dx \, dy = \int_K \lambda^n \, d\lambda,$$

for all $n = 0, \pm 1, \ldots$.

Exercise 1 Show that the mixed spectrum example $T(x, y) = (\alpha + x, x + y) \pmod 1$ (α irrational) has a spectrum which can be described as follows: $U_T | V_d$ has simple discrete spectrum concentrated on the points $\{\exp(2\pi i n\alpha) : n = 0, \pm 1, \ldots\}$ of the circle; $U_T | V_c$ has *countable Lebesgue spectrum*, i.e. there exist $f_1, f_2, \ldots$ (countably infinite) such that $\{U_T^m f_n : m = 0, \pm 1, \ldots ; n = 1, 2, \ldots\}$ is an orthogonal basis for V_c.

Exercise 2 If T_1, T_2 are ergodic measure-preserving transformations of $(X_1, \mathscr{B}_1, m_1)$ and $(X_2, \mathscr{B}_2, m_2)$ respectively, show that the transformation $T_1 \times T_2$ of $(X_1 \times X_2, \mathscr{B}_1 \times \mathscr{B}_2, m_1 \times m_2)$ is ergodic if and only if $H(T_1) \cap H(T_2) = \{1\}$.

Exercise 3 Show that $Tx = 2x \pmod 1$ is ergodic by showing (using Fourier series) that the only T-invariant $L^2[0, 1)$ functions are the constants.

2 Weak-mixing

A measure-preserving transformation of a probability space $(X, \mathscr{B}, m)$ is said to be *weak-mixing* if

$$\frac{1}{N} \sum_{n=0}^{N-1} |m(T^{-n}A \cap B) - m(A) \cdot m(B)| \to 0 \qquad (3.5)$$

for all $A, B \in \mathscr{B}$. Clearly weak-mixing implies ergodicity.

It is easy to deduce from (3.5),

$$\frac{1}{N} \sum_{n=0}^{N-1} |\int (f \circ T^n) \bar{g} \, dm - \int f \, dm \int \bar{g} \, dm| \to 0 \qquad (3.6)$$

for all $f, g \in L^2(X, \mathscr{B}, m)$, by first proving it for simple functions f, g and then approximating arbitrary L^2 functions by simple functions. From (3.6) it follows that weak-mixing is a spectral property.

The following theorem is proved in an analogous manner to the proof of Theorem 1. We therefore omit the proof.

2. Theorem *In order that T should be weak-mixing it is necessary and sufficient that*

$$\frac{1}{N} \sum_{n=0}^{N-1} |\int (f \circ T^n) \bar{f} \, dm| \to 0$$

for all $f \in L^2(X, \mathscr{B}, m)$ with $\int f \, dm = 0$

The following analytic fact (cf. Jacobs [1], p. 79; or Walters [1], p. 40) is useful in connection with weak-mixing.

3. Lemma *If $\{a_n\}$ is a bounded sequence of non-negative numbers then $(1/N) \sum_{n=0}^{N-1} a_n \to 0$ if and only if for some sequence $\mathscr{N}$ of integers of zero density* $\lim_{n \to \infty / n \notin \mathscr{N}} a_n = 0$. *(To say that $\mathscr{N}$ has density zero means $(1/N)$ Card $\mathscr{N} \cap [1, \ldots, N] \to 0$.)*

4. Theorem *The following are equivalent*:

(i) *T is weak-mixing*;

(ii) $T \times T$ *(on $(X \times X, \mathscr{B} \times \mathscr{B}, m \times m)$) is weak-mixing or merely ergodic*;

(iii) U_T *has continuous spectrum in the orthocomplement of the constant functions, i.e. the only eigenfunctions are constant functions.*

Proof Suppose T is weak-mixing. To prove that $T \times T$ is weak-mixing it suffices to prove that

$$\frac{1}{N} \sum_{n=0}^{N-1} |m \times m[(T \times T)^{-n}(A_1 \times A_2) \cap (B_1 \times B_2)]$$
$$- m \times m(A_1 \times A_2) m \times m(B_1 \times B_2)| \to 0$$

for measurable rectangles $A_1 \times A_2$, $B_1 \times B_2$. But we know that

$$\lim_{\substack{n \to \infty \\ n \notin \mathscr{N}_1}} m(T^{-n}A_1 \cap B_1) = m(A_1)m(B_1)$$

and

$$\lim_{\substack{n \to \infty \\ n \notin \mathscr{N}_2}} m(T^{-n}A_2 \cap B_2) = m(A_2)m(B_2)$$

where $\mathscr{N}_1, \mathscr{N}_2$ are sequences of density zero. Hence

$$\lim_{\substack{n \to \infty \\ n \notin \mathscr{N}_1 \cup \mathscr{N}_2}} m \times m[(T \times T)^{-n}(A_1 \times A_2) \cap (B_1 \times B_2)]$$
$$= m \times m(A_1 \times A_2) m \times m(B_1 \times B_2)$$

and $\mathscr{N}_1 \cup \mathscr{N}_2$ has density zero. This proves that (i) implies (ii).

(ii) implies (iii), for if $f(Tx) = \alpha f(x)$, $f \in L^2(X, \mathscr{B}, m)$ and f is non-constant, then $g(Tx, Ty) = g(x, y)$ where $g(x, y) = f(x)\overline{f(y)}$ and g is non-constant.

(iii) implies (i). Let $f \in L^2(X, \mathscr{B}, m)$, $\int f \, dm = 0$; we wish to prove $(1/N) \sum_{n=0}^{N-1} |\int (f \circ T^n) \bar{f} \, dm| \to 0$ or, in view of the characterisation of 'strong-cesaro convergence' already mentioned, (Lemma 3),

$$\frac{1}{N} \sum_{n=0}^{N-1} |\int (f \circ T^n) \bar{f} \, dm|^2 \to 0.$$

(Clearly, $\lim_{n \notin \mathscr{N}} a_n = 0$ if and only if $\lim_{n \notin \mathscr{N}} a_n^2 = 0$ when $\mathscr{N}$ is a sequence of density zero.) But

$$\frac{1}{N}\sum_{n=0}^{N-1}|\int (f\circ T^n)\bar{f}\,dm|^2=\frac{1}{N}\sum_{n=0}^{N-1}|\int_K\lambda^n d\tilde{f}|^2$$

$$=\frac{1}{N}\sum_{n=0}^{N-1}\int_K\lambda^n d\tilde{f}\cdot\int_K\bar{\mu}^n d\tilde{f}$$

$$=\frac{1}{N}\cdot\sum_{n=0}^{N-1}\int\int_{K\times K}(\lambda\bar{\mu})^n d(\tilde{f}\times\tilde{f})$$

$$=\frac{1}{N}\int\int_{K\times K}\frac{(\lambda\bar{\mu})^N-1)}{(\lambda\bar{\mu}-1)}d(\tilde{f}\times\tilde{f})$$

since $\lambda\bar{\mu}=1$ only when $\lambda=\mu$ and $\tilde{f}\times\tilde{f}$ assigns measure zero to the diagonal; for by assumption $\tilde{f}$ is a continuous measure. Obviously

$$\frac{1}{N}\frac{(\lambda\bar{\mu})^N-1}{(\lambda\bar{\mu}-1)}\to 0$$

for (λ,μ) off the diagonal and this convergence is dominated. By Lebesgue's dominated convergence theorem

$$\frac{1}{N}\int\int_{K\times K}\frac{(\lambda\bar{\mu})^N-1}{(\lambda\bar{\mu}-1)}d(\tilde{f}\times\tilde{f})\to 0$$

and the proof of the theorem is complete.

***Exercise* 4** Show that $Tx=2x$ (mod 1) is weak-mixing by showing that the only eigenfunctions in $L^2[0, 1)$ are the constants.

***Exercise* 5** Give examples of measure-preserving transformations which are ergodic but not weak-mixing.

***Exercise* 6** If T_1, T_2 are ergodic and weak-mixing respectively, show that $T_1\times T_2$ is ergodic. Show that it is not weak-mixing unless T_1 is weak-mixing.

3 Strong-mixing

A measure-preserving transformation T of a probability space $(X,\mathscr{B},m)$ is called *strong-mixing* if

$$m(T^{-n}A\cap B)\to m(A)\cdot m(B) \tag{3.7}$$

for all $A,B\in\mathscr{B}$.

In view of what has been said about various equivalent characteris-ations of ergodicity and of weak-mixing, it is easy to show that T is strong-mixing if and only if either of the following holds:

$$\int (f \circ T^n) \bar{g} \, dm \to \int f \, dm \cdot \int \bar{g} \, dm = \langle f, 1 \rangle \langle 1, g \rangle \qquad (3.8)$$

for all $f, g \in L^2(X, \mathcal{B}, m)$;

$$\int (f \circ T^n) \bar{f} \, dm \to 0 \qquad (3.9)$$

for all $f \in L^2(X, \mathcal{B}, m)$ with $\int f \, dm = 0$.

From (3.8) or (3.9) it follows that strong-mixing is a spectral property. Obviously strong-mixing implies weak-mixing (which in turn, as we have remarked, implies ergodicity). Examples of weak-mixing transformations which are not strong-mixing are not easily produced. We shall exhibit one in Chapter 5. For the rest of this section we shall discuss various notions which imply strong-mixing.

If T is a measure-preserving transformation with absolutely continuous spectrum in the orthocomplement of the constants then T is strong-mixing. The assumption here is that for all $f \in L^2(X, \mathcal{B}, m)$ with $\int f \, dm = 0$, $\tilde{f}$ is absolutely continuous with respect to Lebesgue measure on the circle. In this case

$$\int (f \circ T^n) \bar{f} \, dm = \int_K \lambda^n \, d\tilde{f} = \int_K \lambda^n g(\lambda) \, d\lambda \to 0$$

for some $g \in L^1(K, \lambda)$; convergence to zero is a statement of the Riemann – Lebesgue lemma. For the examples known to ergodic theory to this date, absolutely continuous spectrum (in the ortho-complement of the constants) occurs only when U_T has infinite Lebesgue spectrum (in the orthocomplement of the constants), i.e. when there exists $\{f_j : j \in J\}$ (J an infinite set) such that $\{U_T^n f_j : j \in J, n = 0, \pm 1, \dots \}$ is an orthonormal basis $\mathbb{C}^\perp$, the orthocomplement of the constants. A theorem or counter-example in connection with the indicated conjecture would be extremely interesting.

A measure-preserving transformation T of a probability space is said to be a *K-automorphism* (after Kolmogorov) if it is invertible and if there exists a sub-σ-algebra $\mathcal{A}$ such that $T^{-1} \mathcal{A} \subset \mathcal{A}$, $T^n \mathcal{A} \uparrow \mathcal{B}$ (i.e. $\bigcup_n T^n \mathcal{A}$ generates $\mathcal{B}$), and $T^{-n} \mathcal{A} \downarrow \mathcal{N}$ (i.e. $\bigcap_n T^{-n} \mathcal{A} = \mathcal{N}$). As usual $\mathcal{N}$ denotes the trivial σ-algebra.

It is clear that in this case $L^2(X, \mathcal{B}, m)$ can be written

$$L^2(X, \mathcal{B}, m) = \mathbb{C} \oplus \sum_{n=-\infty}^{\infty} U_T^n V,$$

where

$$V = L^2(X, \mathcal{A}, m) \ominus L^2(X, T^{-1}\mathcal{A}, m)$$
$$= \{f : f \in L^2(X, \mathcal{A}, m), \ f \perp L^2(X, T^{-1}\mathcal{A}, m)\}.$$

Taking $\{f_j : j \in J\}$ as an orthonormal basis for V, we get $\{U_T^n f_j : j \in J,$ $n = 0, \pm 1, \ldots \}$ as an orthonormal basis of the orthogonal complement of the constants. Then the spectral properties are fully described when we know the dimension of V; for then U_T will have Lebesgue spectrum with this multiplicity in $\mathbb{C}^\perp$. In fact dim $V = \infty$, and when $L^2(X, \mathcal{B}, m)$ (or $(X, \mathcal{B}, m)$) is separable, this infinity is countable. (Here we assume $(X, \mathcal{B}, m)$ is non-trivial i.e. X is not an atom.)

5. Lemma (Rohlin [4]) *If* $(X, \mathcal{B}, m)$ *is a probability space and if* $\mathcal{C}$ *is a proper non-atomic sub-σ-algebra then* $L^2(X, \mathcal{B}, m) \ominus L^2(X, \mathcal{C}, m)$ *is infinite-dimensional.*

Proof Since $\mathcal{C}$ is proper there exists $f \in L^2(X, \mathcal{B}, m)$ with $0 \neq f \perp L^2(X, \mathcal{C}, m)$. Let $F = \{x : f(x) \neq 0\}$ where $m(F) > 0$. In an obvious sense

$$L^2(F, \mathcal{B} \cap F, m) \ominus L^2(F, \mathcal{C} \cap F, m) \subset L^2(X, \mathcal{B}, m) \ominus L^2(X, \mathcal{C}, m)$$

so we shall show that the left-hand side is infinite-dimensional. (The notation $\mathcal{C} \cap F$ stands for the set of $C \cap F$ with $C \in \mathcal{C}$.) Since $\mathcal{C}$ is non-atomic there exist infinitely many mutually disjoint sets $\{G_i : i = 1, 2, \ldots\}$, $G_i \in \mathcal{C} \cap F$. Defining $g_i = \chi_{G_i}$, we see now that $\{fg_j : j = 1, 2, \ldots\} \subset L^2(F, \mathcal{B} \cap F, m) \ominus L^2(F, \mathcal{C} \cap F, m)$ and $\{fg_j : j = 1, 2, \ldots \}$ is an infinite linearly independent set. The proof is complete.

6. Theorem *If* T *is a K-automorphism then* U_T *has infinite Lebesgue spectrum in* $\mathbb{C}^\perp$.

Proof See the remarks preceding Lemma 5. For the sub-σ-algebra $\mathcal{A}$ with $T^{-1}\mathcal{A} \subset \mathcal{A}$, $T^n \mathcal{A} \uparrow \mathcal{B}$, $T^{-n}\mathcal{A} \downarrow \mathcal{N}$, we have to show that $V = L^2(X, \mathcal{A}, m) \ominus L^2(X, T^{-1}\mathcal{A}, m)$ is infinite-dimensional. This follows from Lemma 5 (with $\mathcal{A} = \mathcal{B}$, $T^{-1}\mathcal{A} = \mathcal{C}$) since $T^{-1}\mathcal{A}$ cannot be atomic (unless $(X, \mathcal{B}, m)$ is trivial). For if $A \in T^{-1}\mathcal{A}$ is an atom then $m(A \cap T^{-n}A) > 0$ for some n and hence $A = T^{-n}A$. Therfore $A = T^{-n}A = T^{-2n}A = \ldots$. We conclude that $A \in \bigcap_n T^{-n}\mathcal{A} = \mathcal{N}$, i.e. $m(A) = 1$. Since $T^n \mathcal{A} \uparrow \mathcal{B}$ and $\mathcal{A}$ is trivial it follows, against our implicit assumption, that $(X, \mathcal{B}, m)$ is trivial.

Corollary *If T is a K-automorphism then T is strong-mixing.*
   Proof We have already remarked that absolutely continuous
spectrum in $\mathbb{C}^\perp$ implies strong-mixing.

T is said to be an *exact endomorphism* if $\bigcap_n T^{-n}\mathcal{B} = \mathcal{N}$. (Of course,
unless $(X, \mathcal{B}, m)$ is trivial, T will not be invertible.)
   In the same way as for K-automorphisms one can prove:

7. Theorem *If T is an exact endomorphism then U_T has infinite 'semi-
Lebesgue' spectrum in the orthocomplement of the constants, i.e. there
exists an infinite orthonormal set $\{f_j : j \in J\}$ of $L^2(X, \mathcal{B}, m)$ functions
such that $\{U_T^n f_j : j \in J, n = 0, 1, \dots\}$ is an orthonormal basis for $\mathbb{C}^\perp$.
Consequently T is strong-mixing.*

A measure-preserving transformation is said to be *mixing of order* 2 if
for all $A, B, C \in \mathcal{B}$

$$m(A \cap T^{-l_n}B \cap T^{-k_n}C) \to m(A) \cdot m(B) \cdot m(C)$$

whenever $\{l_n, k_n\}$ satisfy $l_n \to \infty$, $k_n - l_n \to \infty$. Mixing of higher orders
can be defined similarly. Rohlin [5] has shown that K-
automorphisms and exact endomorphisms are mixing of all orders.
(Order 1 is the same as strong-mixing.) It is possible that strong-
mixing implies mixing of order 2 or, for that matter, mixing of all
orders. No theorems or counter-examples in this connection have
been found to date.

Exercise 7 Show that $Tx = 2x \pmod 1$ is strong-mixing by proving
that $\int \gamma(T^n x)\overline{\gamma(x)}\,dm \to 0$ for all exponential functions $\gamma(x)$
$= \exp(2\pi ikx)$, $k \neq 0$.

4 Markov and Bernoulli shifts

This section relies, to a large extent, on material from the theory of
Markov chains, which can be found in most texts on probability
theory and stochastic processes (cf. for example Chung [2], Feller [1]
or Doob [1]).
   Let P be an irreducible $k \times k$ stochastic matrix, i.e. P is a non-

negative matrix with row sums equal to one and for each i, j there exists n with $P^n(i, j) > 0$. Let

$$X = \prod_{n=-\infty}^{\infty} \{1, 2, \ldots, k\} \quad \text{and} \quad X' = \prod_{n=0}^{\infty} \{1, 2, \ldots, k\},$$

the spaces of doubly infinite and singly infinite sequences of symbols chosen from $1, 2, \ldots, k$. A set of the form $[i_0, \ldots, i_l] = \{x : x_0 = i_0, \ldots, x_l = i_l\}$ is called a cylinder. $T : X \to X$, $T' : X' \to X'$, where $T\{x_n\} = \{x_{n+1}\}$, $T'\{x_n\} = \{x_{n+1}\}$ are called the *two-sided* and *one-sided shifts*, respectively. Let $\mathscr{B}$, $\mathscr{B}'$ be the smallest σ-algebras containing all cylinders such that $T^{-1}\mathscr{B} = \mathscr{B}$, $T'^{-1}\mathscr{B}' \subset \mathscr{B}'$. There is a unique positive left vector $\Lambda = (\lambda_1, \ldots, \lambda_k)$ such that $\Lambda P = \Lambda$ and $\lambda_1 + \cdots + \lambda_k = 1$. Define, for each cylinder,

$m[i_0, \ldots, i_l] = \lambda_{i_0} P(i_0, i_1) \ldots P(i_{l-1}, i_l)$ (or m' for cylinders of X')

then m, m' extend to unique probabilities which we also denote by m, m' such that $m(T^{-1}B) = m(B)$, $m'(T'^{-1}B) = m'(B)$ for all $B \in \mathscr{B}$, $B \in \mathscr{B}'$ respectively. $(X, \mathscr{B}, m, T)$ and $(X', \mathscr{B}', m', T')$ (or even T and T') are called the two-sided and one-sided *Markov chains* or *shifts* (*Markov automorphisms* and *endomorphisms*) defined by P.

Let d be the highest common factor of $\{n : P^n(i, i) > 0\}$. Then d is independent of i and is called the *period* of P. P is said to be *aperiodic* if $d = 1$. For each $i = 1, \ldots, k$ there exists N such that $P^{nd}(i, i) > 0$ whenever $n \geq N$ and for all n not divisible by d, $P^n(i, i) = 0$.

Moreover the symbols $1, 2, \ldots, k$ can be divided into d cyclically moving classes $S_0, S_1, S_2, \ldots, S_{d-1}$ so that $P_{i,j} > 0$ only if $i \in S_k$, $j \in S_{k'}$ where $k' = k + 1 \pmod{d}$.

If each row of P is identical then T, T' are called *Bernoulli* shifts (*Bernoulli automorphisms and endomorphisms* respectively). In this case

$$P = \begin{pmatrix} \Lambda \\ \vdots \\ \Lambda \end{pmatrix} \quad \Lambda P = \Lambda,$$

and the probability spaces on which T, T' act are infinite direct products of the finite probability space $(1, 2, \ldots, k)$ with measures $\lambda_1, \ldots, \lambda_k$. Of course, P in this case is aperiodic.

If P is an aperiodic irreducible stochastic $k \times k$ matrix then $P^n(i,j) \to \lambda_j$, i.e.

$$P^n \to \begin{pmatrix} \Lambda \\ \vdots \\ \Lambda \end{pmatrix}.$$

8. Theorem *An aperiodic Markov automorphism is a K-automorphism. An aperiodic Markov endomorphism is an exact endomorphism.*

Proof If T is a two-sided Markov shift defined on the doubly infinite direct product $X = \prod_{n=-\infty}^{\infty} \{1, 2, \ldots, k\}$, let $\mathscr{A}$ denote the σ-algebra generated by cylinders $[i_0, \ldots, i_l] = \{x : x_j = i_j, 0 \leq j \leq l\}$, $l = 0, 1, \ldots$. It is clear that $T^n \mathscr{A} \uparrow \mathscr{B}$, the smallest σ-algebra containing cylinders which is strictly T-invariant. Therefore it suffices to show that $\bigcap_n T^{-n} \mathscr{A} \downarrow \mathscr{N}$. Evidently this is the same as showing that the one-sided Markov shift T' is an exact endomorphism. Let m' be defined by the irreducible aperiodic $k \times k$ matrix P. Define $C = [i_0, \ldots, i_l]$, $D = [j_0, \ldots, j_m]$ then

$$m'(C \cap T'^{-n}D) = \sum_{x_0, \ldots, x_{n-l}} m'[i_0, \ldots, i_l, x_0, \ldots, x_{n-l-1}, j_0, \ldots, j_m]$$

$$= \sum_{x_0, \ldots, x_{n-l}} \lambda_{i_0} P(i_0, i_1) \ldots P(j_{m-1}, j_m)$$

$$= \lambda_{i_0} P(i_0, i_1) \ldots P^{n-l}(i_l, j_0) P(j_0, j_1) \ldots P(j_{m-1}, j_m)$$

$$= m'(C) m'(D) [P^{n-l}(i_l, j_0)/\lambda_{j_0}] \quad \text{(for } n \geq l\text{)}.$$

As we have remarked this last bracket converges to 1. Hence given a cylinder C then for $1 > \varepsilon > 0$ there exists N such that

$$(1 - \varepsilon) m'(C) m'(D) \leq m'(C \cap T'^{-n}D) \leq (1 + \varepsilon) m'(C) m'(D)$$

for all cylinders D whenever $n > N$. Hence

$$(1 - \varepsilon) m'(C) \leq E(\chi_C | T'^{-n} \mathscr{B}'_m) \leq (1 + \varepsilon) m'(C)$$

for large enough n, where $\mathscr{B}'_m$ is the σ-algebra generated by the cylinders of length m. Letting $m \to \infty$ and applying the Increasing Martingale theorem, with n fixed, we have

$$(1 - \varepsilon) m'(C) \leq E(\chi_C | T'^{-n} \mathscr{B}') \leq (1 + \varepsilon) m'(C).$$

Now let $n \to \infty$ and apply the Decreasing Martingale theorem. Then for $\mathscr{B}'_\infty = \bigcap_n T'^{-n} \mathscr{B}'$ we have

$$(1 - \varepsilon)m'(C) \leq E(\chi_C | \mathscr{B}'_\infty) \leq (1 + \varepsilon)m'(C).$$

Let $A \in \mathscr{B}'_\infty$ and integrate this inequality over A to obtain

$$(1 - \varepsilon)m'(C)m'(A) \leq m'(A \cap C) \leq (1 + \varepsilon)m'(C)m'(A).$$

$1 > \varepsilon > 0$ was arbitrary so we may conclude that for each $A \in \mathscr{B}'_\infty$, and cylinder C (or finite union of cylinders)

$$m'(A \cap C) = m'(A) \cdot m'(C).$$

Hence $m'(A)m'(C) = m'(A \cap C)$ for all $A \in \mathscr{B}'_\infty$ and all $C \in \mathscr{B}'_n$, i.e. $E(\chi_A | \mathscr{B}'_n) = m'(A)$ and taking limits, using the Increasing Martingale theorem, $\chi_A = m'(A)$ a.e. In other words $m'(A) = 0$ or 1. Therefore $\mathscr{B}'_\infty$ is trivial.

Exercise 8   Show that $Tx = 2x \pmod 1$ is an exact endomorphism by using the dyadic intervals

$$I_n(y) = \left\{ x : \frac{y_0}{2} + \frac{y_1}{2^2} + \cdots + \frac{y_n}{2^{n+1}} \leq x \leq \frac{y_0}{2} + \cdots + \frac{y_n + 1}{2^{n+1}} \right\}$$

$(y_n = 0, y_i = 0, 1)$ in the same way as the cylinder sets used in this section.

4
Entropy

---

1 The isomorphism problem

If T_1, T_2 are measure-preserving transformations of probability spaces $(X_1, \mathscr{B}_1, m_1), (X_2, \mathscr{B}_2, m_2)$ respectively, we say that T_2 is a *factor* of T_1 or that T_2 is a *homomorphic image* of T_1, if there is a measure-preserving transformation φ of almost all X_1 onto almost all X_2 such that $\varphi T_1 = T_2 \varphi$ a.e. The fact that φ is measure-preserving is sometimes written $\varphi : (X_1, \mathscr{B}_1, m_1) \to (X_2, \mathscr{B}_2, m_2)$ or $(X_1, \mathscr{B}_1, m_1) \xrightarrow{\varphi} (X_2, \mathscr{B}_2, m_2)$. '$T_2$ is a homomorphic image of T_1 (by the homomorphism φ)' is abbreviated by $T_1 \xrightarrow{\varphi} T_2$.

If $T_1 \xrightarrow{\varphi} T_2$ and $T_2 \xrightarrow{\psi} T_1$ we say that T_1 and T_2 are *weakly isomorphic*. If in addition $\varphi \circ \psi = $ identity a.e. and $\psi \circ \varphi = $ identity a.e. then T_1, T_2 are said to be *isomorphic* $(T_1 \simeq T_2)$. It is clear that weak isomorphism and isomorphism are equivalence relations on the collection of all measure-preserving transformations.

If T_1 and T_2 are isomorphic then the invertible isometry induced by the isomorphism gives rise to spectral equivalence between U_{T_1} and U_{T_2}. Thus spectral invariants should be regarded as the primary invariants of a measure-preserving transformation. The isomorphism problem of ergodic theory can be formulated as follows. Given two measure-preserving transformations T_1, T_2 such that U_{T_1}, U_{T_2} are spectrally equivalent, when can we say that T_1 and T_2 are isomorphic? More specifically, can we find non-spectral isomorphism invariants?

Halmos and von Neumann showed that $U_{T_1} \simeq U_{T_2}$ implies $T_1 \simeq T_2$ in the special case that T_1, T_2 are ergodic with purely discrete spectrum acting on the unit interval with Lebesgue measure (or probability spaces isomorphic to the unit interval). We shall not prove this theorem. The interested reader is referred for details to von Neumann [3]; Halmos and von Neumann [1]. The Halmos and von Neumann classification theory is summarised in the following:

Discrete spectrum classification Theorem *If T_1, T_2 are ergodic measure-preserving transformations of the unit interval with discrete spectrum then $T_1 \simeq T_2$ if and only if $H(T_1) \equiv H(T_2)$. For every countable subgroup $H \subset K = \{z \in \mathbb{C} : |z| = 1\}$ there exists an ergodic measure-preserving transformation T with discrete spectrum such that $H(T) = H$. Every ergodic measure-preserving transformation with discrete spectrum (acting on the unit interval) is isomorphic to a translation of a compact metric abelian group.*

Remark 1 We stress that the unit interval assumptions above are not nearly as restrictive as they appear; in fact one has to make special pathological efforts to construct non-atomic probability spaces which are not isomorphic to the unit interval.

Remark 2 Any non-atomic probability space obtained as the completion of a Borel measure on a complete separable metric space is isomorphic to the unit interval with Lebesgue sets and measure (Halmos and von Neumann [1]; Rohlin [1]). The Bernoulli and Markov sequence spaces of Chapter 3, § 4, are thus isomorphic to the unit interval.

Remark 3 The most important property of the unit interval probability space is that an algebraic *set* isomorphism (defined on sets modulo null sets) can always be realised by a measure-preserving *point* transformation.

It is not difficult to produce ergodic examples T_1, T_2 with mixed spectrum such that $U_{T_1} \simeq U_{T_2}$ but $T_1 \not\simeq T_2$. In fact Abramov [1] systemised a class of such examples and generalised the discrete spectrum classification theorem, by using the notion of quasi-discrete spectrum (which is only partially a spectral notion). His invariants are a sequence of groups and group homomorphisms. The class of nilflows and unipotent affine transformations which act on compact nilmanifolds are closely related to Abramov's class and have been studied and classified in Auslander, Green and Hahn [1]; and Parry [1], [2]. All of these transformations have countable Lebesgue spectrum in V_c.

The problem of classifying ergodic measure-preserving transformations with continuous spectrum (in the orthocomplement of the constants) remained open for many years. Kolmogorov [3] introduced the concept of entropy (from information theory) to produce non-isomorphic examples with the same continuous spectrum. To do this he showed that entropy is an isomorphism invariant and showed that it is readily computable for Bernoulli automorphisms. As we have seen the Bernoulli automorphisms, being K-automorphisms, are all spectrally isomorphic since they have countable Lebesgue spectrum in the orthocomplement of the constants.

Thus Kolmogorov produced a continuum of non-isomorphic measure-preserving transformations all of which are spectrally isomorphic (with continuous spectrum) since entropy is a numerical invariant which, when restricted to the Bernoulli automorphisms, has the positive real line for its range.

Not surprisingly, after this result, there was a concerted attack on the classification problem using the new invariant. Initially the major work was done by the Russian school – Kolmogorov, Rohlin, Sinai and others. But later Ornstein made a major breakthrough complementing Kolmogorov's result. Ornstein [3] showed that two Bernoulli automorphisms with the same entropy are isomorphic. (Sinai [4] had come close to this when he showed that they are weakly isomorphic.) This result was later extended by Friedman and Ornstein [2] to include aperiodic Markov automorphisms. The natural conjecture before, and especially, after, these results was that equal entropies would imply isomorphism for K-automorphisms. However, Ornstein [4] demolished this by constructing K-automorphisms without square roots. Ornstein and Shields [1] went on to produce uncountably many non-isomorphic K-automorphisms with the same entropy. The main developments since these results have been concerned with:

(*a*) Extending these results to Bernoulli flows (Ornstein [5], [6], [7]);

(*b*) Showing that a variety of examples are isomorphic to Bernoulli automorphisms or flows (Adler and Weiss [1]; Katznelson [1]; Smorodinsky [1]; Wilkinson [1]; Rudolfer and Wilkinson [1]; Miles and Thomas [1]; Sinai [5]; Bowen [1]; Dani [1]);

(*c*) Relating these results to statistical mechanics (Bowen [2]; Walters [2]; Ruelle [1]; Gallovotti [1]; Sinai [6]);

(*d*) Investigating K-flows and Bernoulli flows under changes of velocity (Feldman [1]; Weiss [2]).

There are now (or will be soon) several good accounts of 'Bernoulli theory' (Shields [1]; Weiss [1]; Friedman and Ornstein [1]; Ornstein [2]; and the seminar notes of Moser, Phillips and Varadhan [1]). We have mentioned developments in Bernoulli theory because of their importance but we shall only provide an account of some of the earlier entropy theory – first steps as it were (cf. also Billingsley [1]; Parry [3]; Rohlin [4]).

2 Entropy as an invariant

We refer the reader to Chapter 2 for the definitions of information and entropy and the proofs of their most elementary identities and inequalities. We shall assume throughout the remainder of this chapter that $(X, \mathcal{B}, m)$ is a separable probability space.

If $(X, \mathcal{B}, m)$ is a separable probability space, and α, $\beta \le \gamma$, are countable partitions, then for any sub-σ-algebras $\mathcal{A}_1 \subset \mathcal{A}_2$ we have:

$$H(\beta|\mathcal{A}_1) \le H(\gamma|\mathcal{A}_1),$$
$$H(\beta|\mathcal{A}_2) \le H(\beta|\mathcal{A}_1),$$

and

$$H(\alpha \vee \beta|\mathcal{A}_1) = H(\alpha|\mathcal{A}_1) + H(\beta|\hat{\alpha} \vee \mathcal{A}_1)$$
$$\le H(\alpha|\mathcal{A}_1) + H(\beta|\mathcal{A}_1).$$

If T is a measure-preserving transformation then

$$H(\beta|\mathcal{A}_1) = H(T^{-1}\beta|T^{-1}\mathcal{A}_1).$$

When $\mathcal{A}$ is a sub-σ-algebra $\mathcal{A}^-$ will denote $\bigvee_{i=1}^{\infty} T^{-i}\mathcal{A}$ so that $T^{-1}(\mathcal{A} \vee \mathcal{A}^-) = \mathcal{A}^- \subset (\mathcal{A} \vee \mathcal{A}^-)$. If T is an *invertible* measure-preserving transformation, $\mathcal{A}_T$ will denote $\bigvee_{i=-\infty}^{\infty} T^i\mathcal{A}$ so that $T^{-1}\mathcal{A}_T = \mathcal{A}_T$. A countable partition ξ is called a *strong-generator* for T when $\hat{\xi} \vee \hat{\xi}^- = \mathcal{B}$ and a *generator* (if T is invertible) when $\hat{\xi}_T = \mathcal{B}$.

If ξ is a countable partition with finite entropy then

$$h(T, \xi) = \lim_{N \to \infty} \frac{1}{N+1} H(\xi \vee T^{-1}\xi \vee \cdots \vee T^{-N}\xi) = H(\xi|\hat{\xi}^-).$$

The entropy of a measure-preserving transformation T (referred to in §1, this Chapter) is defined as

$$h(T) = \sup_{H(\xi) < \infty} h(T, \xi).$$

It is clear that $h(T)$ is an isomorphism (in fact weak isomorphism) invariant; for if S is a factor of T ($T \overset{\varphi}{\to} S$) then $h(S, \xi) = h(T, \varphi^{-1}\xi)$ and $h(S) \leq h(T)$, and similarly if T is a factor of S, $h(T) \leq h(S)$.

$h(T)$ is sometimes referred to as the Kolmogorov – Sinai invariant after Sinai's modifications of the entropy invariant introduced by Kolmogorov [3]. The point to be stressed is that $h(T)$ is frequently computable and distinguishes between many measure-preserving transformations with identical continuous (even countable Lebesgue) spectrum in the orthogonal complement of the constants.

Notice that it is immediate from the definition that $h(T) = 0$ if and only if $H(\xi|\hat{\xi}^-) = 0$ for all ξ with $H(\xi) < \infty$, i.e. if and only if $\hat{\xi} \subset \hat{\xi}^-$ for all ξ with $H(\xi) < \infty$. The phenomenon of the 'present' $\hat{\xi}$ being contained in the 'future' $\hat{\xi}^-$ (or 'past') is known as *determinism*. Thus a zero entropy transformation is *completely deterministic*.

If T has zero entropy then $T^{-1} \mathscr{B} = \mathscr{B}$.

For if $T^{-1} \mathscr{B} \subset \mathscr{B}$ (properly) then there exists $B \in \mathscr{B}$, $B \notin T^{-1} \mathscr{B}$, and taking $\xi = (B, B^c)$ we have $H(\xi|T^{-1}\mathscr{B}) \neq 0$, and hence $H(\xi|\hat{\xi}^-) \neq 0$, i.e. $h(T) > 0$.

1. Theorem *If T is a measure-preserving transformation then for each positive integer n, $h(T^n) = nh(T)$. If T is an automorphism $h(T) = h(T^{-1})$. (The entropy of the identity and therefore of any measure-preserving transformation T with $T^k = identity$ (some k) is zero.)*

Proof The bracketed statement is trivial. The fact that $h(T) = h(T^{-1})$, for invertible transformations, follows from the identity $H(\bigvee_{n=0}^{N} T^{-n}\xi) = H(\bigvee_{n=0}^{N} T^n \xi)$. Let T be a measure-preserving transformation and let n be a positive integer. For each partition ξ with finite entropy we have

$$h\left(T^n, \bigvee_{i=0}^{n-1} T^{-i}\xi\right) = \lim_{N \to \infty} \frac{1}{N} H\left(\bigvee_{i=0}^{nN-1} T^{-i}\xi\right)$$
$$= nh(T, \xi).$$

Hence $nh(T, \xi) \geq h(T^n, \xi)$ and $nh(T) \geq h(T^n)$. On the other hand, taking suprema over ξ with finite entropy yields $nh(T) \leq h(T^n)$.

2. Lemma *For two partitions ξ, η with finite entropy*

$$h(T, \xi) \le h(T, \eta) + H(\xi|\eta).$$

Proof In fact

$$H\left(\bigvee_{n=0}^{N} T^{-n}\xi\right) \le H\left(\bigvee_{n=0}^{N} T^{-n}\xi \vee \bigvee_{n=0}^{N} T^{-n}\eta\right)$$

$$\le H\left(\bigvee_{n=0}^{N} T^{-n}\eta\right)$$
$$+ H\left(\bigvee_{n=0}^{N} T^{-n}\xi \,\middle|\, \bigvee_{n=0}^{N} T^{-n}\eta\right)$$

$$\le H\left(\bigvee_{n=0}^{N} T^{-n}\eta\right)$$
$$+ \sum_{n=0}^{N} H\left(T^{-n}\xi \,\middle|\, \bigvee_{n=0}^{N} T^{-n}\eta\right)$$

$$\le H\left(\bigvee_{n=0}^{N} T^{-n}\eta\right) + \sum_{n=0}^{N} H(T^{-n}\xi|T^{-n}\eta)$$

$$= H\left(\bigvee_{n=0}^{N} T^{-n}\eta\right) + (N+1)H(\xi|\eta).$$

Dividing by $N + 1$ and taking limits, the proof is concluded.

3. Theorem *If $\hat{\xi}_n \uparrow \mathscr{B}$, $H(\xi_n) < \infty$ then*

$$h(T) = \lim_{n \to \infty} h(T, \xi_n).$$

Proof Let $H(\eta) < \infty$; then $h(T, \eta) \le h(T, \xi_n) + H(\eta|\xi_n)$ and by Theorem 6, Chapter 2, $H(\eta|\xi_n) \to 0$. Hence $h(T, \eta) \le \lim_{n \to \infty} h(T, \xi_n)$ and this is true for all $\eta(H(\eta) < \infty)$.

In the following the direct product of two partitions α, β, denoted by $\alpha \times \beta$ is the partition consisting of sets $A \times B$ with $A \in \alpha$, $B \in \beta$.

Corollary *For any two measure-preserving transformations S, T*

$$h(S \times T) = h(S) + h(T).$$

Proof Let $\mathscr{B}_S$, $\mathscr{B}_T$ be the σ-algebras of the spaces on which S, T act. Choose $\hat{\xi}_n \uparrow \mathscr{B}_S$, $\hat{\eta}_n \uparrow \mathscr{B}_T$, $H(\xi_n) < \infty$, $H(\eta_n) < \infty$. Then

$\hat{\xi}_n \times \hat{\eta}_n \uparrow \mathscr{B}_S \times \mathscr{B}_T$ and it is easy to see that $h(S \times T, \xi_n \times \eta_n) = h(S, \xi_n) + h(T, \eta_n)$ using the additivity of entropy over the refinement of two independent partitions. Taking limits and applying the theorem the proof is concluded.

4. Theorem

(i) *If T is a measure-preserving transformation and ξ, η satisfy $\hat{\eta} \vee \hat{\eta}^- \supset \hat{\xi}$ ($H(\xi) < \infty$, $H(\eta) < \infty$) then*

$$h(T, \eta) \geq h(T, \xi).$$

(ii) *If T is invertible and ξ, η satisfy $\hat{\eta}_T \supset \hat{\xi}$ ($H(\xi) < \infty$, $H(\eta) < \infty$) then*

$$h(T, \eta) \geq h(T, \xi).$$

In particular, if η is a strong-generator or generator (for the invertible case) with finite entropy then

$$h(T, \eta) = h(T).$$

Proof The last statement is an immediate deduction from (i) and (ii). The proofs of (i) and (ii) are similar so we shall prove (ii) only. By Lemma 2 we have

$$h(T, \xi) \leq h\left(T, \bigvee_{i=-n}^{n} T^i \eta\right) + H\left(\xi \middle| \bigvee_{i=-n}^{n} T^i \eta\right)$$

and the last term converges to $H(\xi|\hat{\eta}_T) = 0$. However,

$$h\left(T, \bigvee_{i=-n}^{n} T^i \eta\right) = \lim_{N \to \infty} \frac{1}{(N+1)} H\left(\bigvee_{i=-n}^{N+n} T^{-i}\eta\right)$$

$$= \lim_{N \to \infty} \frac{1}{(N+1)} H\left(\bigvee_{i=0}^{N+2n} T^{-i}\eta\right)$$

$$= h(T, \eta).$$

Hence $h(T, \xi) \leq h(T, \eta)$.

We have now proved Kolmogorov's result:

(i) *All Bernoulli automorphisms (in fact aperiodic Markov automorphisms) have countable Lebesgue spectrum in the orthocomplement of the constants, since they are K-automorphisms (cf. Theorem 6, Chapter 3).*

(ii) *There is a continuum of Bernoulli automorphisms which are non-isomorphic to each other since their entropies cover the positive reals. The entropy of a Bernoulli automorphism generated by $p_1, \ldots, p_k$ is*

$- \sum_i p_i \log p_i$. In particular, the Bernoulli automorphisms $(\frac{1}{2}, \frac{1}{2})$ and $(\frac{1}{3}, \frac{1}{3}, \frac{1}{3})$ are non-isomorphic having entropies $\log 2$, $\log 3$, respectively.

Exercise 1 Show that the transformation $2x \,(\mathrm{mod}\; 1)$ has entropy $\log 2$. Establish an isomorphism between this transformation and the Bernoulli endomorphism generated by $(\frac{1}{2}, \frac{1}{2})$.

Exercise 2 Show that a Markov automorphism (or endomorphism) T defined by the transition matrix P has entropy $h(T) = - \sum_{i,j} \lambda_i P(i,j) \log P(i,j)$ where $\Lambda P = \Lambda$, $\Lambda = (\lambda_1, \ldots, \lambda_k)$, $\lambda_1 + \cdots + \lambda_k = 1$. A Bernoulli automorphism or endomorphism T generated by $(p_1, \ldots, p_k)$ has entropy $h(T) = - \sum_i p_i \log p_i$.

Exercise 3 If T is a measure-preserving transformation of the unit interval with a finite generator ξ such that $h(T) = H(\xi)$, show that T is isomorphic to a Bernoulli automorphism. If ξ is a strong-generator with $h(T) = H(\xi)$ then T is isomorphic to a Bernoulli endomorphism. (It is enough to establish an algebraic set isomorphism between the σ-algebra of the sequence space and the Lebesgue σ-algebra of the unit interval (cf. Remarks 2, 3, §1, and Exercise 5, Chapter 2).)

Exercise 4 With the same hypotheses as in Exercise 3, except that $h(T) = H(\xi | T'^{-1}\xi)$, show that T is isomorphic to a Markov automorphism (endomorphism). (Note that $H(\xi | T^{-1}\xi) = H(\xi | T^{-1}\xi \vee \cdots \vee T^{-n}\xi)$ for all $n \geq 1$. When $n = 2$, for example, use Exercise 6, Chapter 2, to show that

$$\frac{m(A \cap T^{-1}B)}{m(T^{-1}B)} = \frac{m(A \cap T^{-1}B \cap T^{-2}C)}{m(T^{-1}B \cap T^{-2}C)}$$

for all $A, B, C \in \xi$.)

3 The Pinsker σ-algebra

If T is a measure-preserving transformation of the separable probability space $(X, \mathscr{B}, m)$ then

$$\mathscr{R}(T) = \bigcap_n T^{-n}\mathscr{B}$$

is called the *Rohlin σ-algebra*. Evidently $T^{-1} \mathcal{R}(T) = \mathcal{R}(T)$ and, if $\mathcal{A}$ is a σ-algebra contained in $\mathcal{B}$ such that $T^{-1} \mathcal{A} = \mathcal{A}$, then $\mathcal{A} \subset \mathcal{R}(T)$.

The *Pinsker σ-algebra* $\mathcal{P}(T)$ is defined as the smallest σ-algebra containing $\bigcup \{\hat{\xi} : h(T, \xi) = 0, H(\xi) < \infty\}$. This latter collection is an algebra as may be verified using the facts that

(i) $A \in \xi, H(\xi) < \infty, h(T, \xi) = 0$ together imply $(A, A^c) \leq \xi$ and $h(T, \{A, A^c\}) = 0$;

(ii) $h(T, \xi) = 0, \ h(T, \eta) = 0, \ H(\xi) < \infty, \ H(\eta) < \infty$ together imply $h(T, \xi \vee \eta) \leq h(T, \xi) + h(T, \eta) = 0$.

Clearly $T^{-1} \mathcal{P}(T) \subset \mathcal{P}(T)$, but the converse inclusion also holds since $h(T, \xi) = 0$ implies $\hat{\xi} \subset \hat{\xi}^-$ and therefore $\xi = T^{-1} \eta$ for some $\hat{\eta} \subset \mathcal{P}(T)$. Therefore $T^{-1} \mathcal{P}(T) = \mathcal{P}(T)$. $\mathcal{P}(T)$ is the *deterministic part* of $\mathcal{B}$.

The Pinsker σ-algebra may be thought of as the largest σ-algebra 'with' zero entropy. To justify this description we need:

5. Theorem *If $H(\eta) < \infty$, $\hat{\eta} \subset \mathcal{P}(T)$ then $h(T, \eta) = 0$. (The converse is obvious.)*

Proof Since $\bigcup \{\hat{\xi} : h(T, \xi) = 0, H(\xi) < \infty\}$ is an algebra which generates $\mathcal{P}(T)$, there exists $\hat{\xi}_n$ in this algebra, ξ_n finite, with $\hat{\xi}_n \uparrow \mathcal{P}(T)$ (cf. Exercise 11, Chapter 2). But

$$h(T, \eta) \leq h(T, \xi_n) + H(\eta | \hat{\xi}_n) = H(\eta | \hat{\xi}_n).$$

Hence, taking limits we have $h(T, \eta) \leq H(\eta | \mathcal{P}(T)) = 0$.

T is said to have *completely positive entropy* if $h(T, \xi) > 0$ whenever $H(\xi) < \infty$ and ξ is non-trivial, i.e. $\xi \neq \nu = \{X\}$. In other words T has completely positive entropy precisely when $\mathcal{P}(T) = \mathcal{N}$ (the trivial σ-algebra of sets of measure zero and one.)

Our object is to prove Rohlin and Sinai's result (cf. Rohlin and Sinai [1]) to the effect that an invertible transformation is a K-automorphism when and only when it has completely positive entropy.

We shall need the easily proved facts: $\mathcal{P}(T) = \mathcal{P}(T^k)$, $k \geq 1$, and $\mathcal{P}(T) = \mathcal{P}(T^{-1})$ when T is invertible.

Exercise 5 If T is a measure-preserving transformation, prove that

$\mathscr{P}(T) = \mathscr{P}(T^k)$ for all $k \geq 1$. If T is also invertible prove that $\mathscr{P}(T) = \mathscr{P}(T^{-1})$.

4 Rohlin–Sinai theorem

Throughout this section T will be an *invertible* measure-preserving transformation of a separable probability space.

6. Lemma *If* $H(\alpha) < \infty$, $H(\beta) \leq \infty$ *and* $\beta \leq \alpha$ *or* $\alpha \leq \beta$ *then* $(1/n)H(\alpha^n | \hat{\beta}^-) \to H(\alpha | \hat{\alpha}^-)$ *where* $\alpha^n = \alpha \vee \cdots \vee T^n \alpha$.

Proof (i) Suppose $\beta \leq \alpha$. $T^{-n}(\hat{\beta}^- \vee \hat{\alpha}^{n-1}) \uparrow \hat{\alpha}^-$ so that $H(\alpha | T^{-n}(\hat{\beta}^- \vee \hat{\alpha}^{n-1})) \to H(\alpha | \hat{\alpha}^-)$ and

$$H(\alpha^n | \hat{\beta}^-) = H(\alpha | \hat{\beta}^-) + H(T\alpha | \hat{\alpha} \vee \hat{\beta}^-) + \cdots$$
$$+ H(T^n \alpha | \hat{\alpha}^{n-1} \vee \hat{\beta}^-),$$

so that $(1/n)H(\alpha^n | \hat{\beta}^-) \to H(\alpha | \hat{\alpha}^-)$.

(ii) Suppose $\alpha \leq \beta$. Then

$$\frac{1}{n}H(\alpha^n | \hat{\beta}^-) \leq \frac{1}{n}H(\alpha^n | \hat{\alpha}^-) \to H(\alpha | \hat{\alpha}^-) \quad \text{by (i).}$$

Conversely

$$\frac{1}{n}H(\alpha^n | \hat{\beta}^-) = \frac{1}{n}H(\beta^n | \hat{\beta}^-) - \frac{1}{n}H(\beta^n | \hat{\alpha}^n \vee \hat{\beta}^-),$$

and taking limits we have

$$\lim_{n \to \infty} \frac{1}{n}H(\alpha^n | \hat{\beta}^-) \geq H(\beta | \hat{\beta}^-) - \lim_{n \to \infty} \frac{1}{n}H(\beta^n | \hat{\alpha}^n \vee \hat{\alpha}^-)$$
$$= \lim_{n \to \infty} \left[\frac{1}{n}H(\beta^n | \hat{\alpha}^-) - \frac{1}{n}H(\beta^n | \hat{\alpha}^n \vee \hat{\alpha}^-) \right]$$
$$= \lim_{n \to \infty} \frac{1}{n}H(\alpha^n | \hat{\alpha}^-) = H(\alpha | \hat{\alpha}^-).$$

The following will be used in the proof of Theorem 12.

7. Lemma *Suppose* α, β, γ *have finite entropies and* $\alpha \leq \beta$; *then*

$$H(\alpha | \hat{\beta}^- \vee T^{-n}\hat{\gamma}^-) \to H(\alpha | \hat{\beta}^-).$$

Proof $H(\alpha | \hat{\beta}^- \vee T^{-n}\hat{\gamma}^-) = H(\beta | \hat{\beta}^- \vee T^{-n}\hat{\gamma}^-)$
$$- H(\beta | \hat{\alpha} \vee \hat{\beta}^- \vee T^{-n}\hat{\gamma}^-).$$

Now consider the equation

$$H(\beta^n|\hat{\beta}^- \vee \hat{\gamma}^-) = H(\beta|\hat{\beta}^- \vee \hat{\gamma}^-) + H(T\beta|T\hat{\beta}^- \vee \hat{\gamma}^-)$$
$$+ \cdots + H(T^n\beta|T^n\hat{\beta}^- \vee \hat{\gamma}^-)$$
$$= H(\beta|\hat{\beta}^- \vee \hat{\gamma}^-) + H(\beta|\hat{\beta}^- \vee T^{-1}\hat{\gamma}^-)$$
$$+ \cdots + H(\beta|\hat{\beta}^- \vee T^{-n}\hat{\gamma}^-)$$

which yields

$$H(\beta|\hat{\beta}^-) = \lim_{n \to \infty} \frac{1}{n} H(\beta^n|\hat{\beta}^- \vee \hat{\gamma}^-) = \lim_{n \to \infty} H(\beta|\hat{\beta}^- \vee T^{-n}\hat{\gamma}^-)$$

(with Lemma 6). Applying this to our first equation we have

$$\lim_{n \to \infty} H(\alpha|\hat{\beta}^- \vee T^{-n}\hat{\gamma}^-) = H(\beta|\hat{\beta}^-) - \lim_{n \to \infty} H(\beta|\hat{\alpha} \vee \hat{\beta}^- \vee T^{-n}\hat{\gamma}^-).$$

Hence

$$\lim_{n \to \infty} H(\alpha|\hat{\beta}^- \vee T^{-n}\hat{\gamma}^-) \geq H(\beta|\hat{\beta}^-) - H(\beta|\hat{\alpha} \vee \hat{\beta}^-)$$
$$= H(\alpha|\hat{\beta}^-).$$

The reverse inequality is obvious.

8. Theorem *For $H(\alpha) < \infty$, $H(\beta) < \infty$ we have*
$$h(T, \alpha \vee \beta) = h(T, \beta) + H(\alpha|\hat{\beta}_T \vee \hat{\alpha}^-).$$
Proof $(1/n)H(\alpha^n \vee \beta^n|\hat{\alpha}^- \vee \hat{\beta}^-)$
$$= \frac{1}{n}H(\beta^n|\hat{\alpha}^- \vee \hat{\beta}^-) + \frac{1}{n}H(\alpha^n|\hat{\alpha}^- \vee \hat{\beta}^- \vee \beta^n)$$
$$= \frac{1}{n}H(\beta^n|\hat{\alpha}^- \vee \hat{\beta}^-) + \frac{1}{n}\sum_{k=0}^{n} H(\alpha|\hat{\alpha}^- \vee \hat{\beta}^- \vee \beta^k)$$

and taking limits we obtain, by Lemma 6,
$$h(T, \alpha \vee \beta) = h(T, \beta) + H(\alpha|\hat{\beta}_T \vee \hat{\alpha}^-).$$

Lemmas 6, 7 and Theorem 8 can be relativised with respect to a strictly invariant σ-algebra $\mathscr{A}$ $(T^{-1}\mathscr{A} = \mathscr{A})$, i.e. these results remain valid (and the proofs are essentially unaltered) if we refine 'denominators' against $\mathscr{A}$.

9. Theorem $\mathscr{P}(T) = \bigvee_{H(\xi) < \infty} \bigcap_n T^{-n}\hat{\xi}^-$

Proof Choose $\hat{\xi}_k \uparrow \mathscr{P}(T)(H(\xi_k) < \infty)$; then

$$\hat{\xi}_k \subset \hat{\xi}_k^- = \bigcap_n T^{-n} \hat{\xi}_k^- \subset \mathscr{P}(T),$$

i.e.

$$\mathscr{P}(T) = \bigvee_k \bigcap_n T^{-n} \hat{\xi}_k^- \subset \bigvee_{H(\xi) < \infty} \bigcap_n T^{-n} \hat{\xi}^-.$$

On the other hand, if $\hat{\eta} \subset \bigcap_n T^{-n} \hat{\xi}^-$, then $\hat{\eta}_T \subset \hat{\xi}^-$ and

$$H(\xi \vee \eta | \hat{\xi}^- \vee \hat{\eta}^-) = H(\xi | \hat{\xi}^-) = H(\xi | \hat{\xi}^- \vee \hat{\eta}_T) + H(\eta | \hat{\eta}^-)$$
$$= H(\xi | \hat{\xi}^-) + H(\eta | \hat{\eta}^-).$$

Hence $H(\eta | \hat{\eta}^-) = 0$ so that $\hat{\eta} \subset \mathscr{P}(T)$. Therefore $\mathscr{P}(T) \supset \bigcap_n T^{-n} \hat{\xi}^-$ whenever $H(\xi) \leq \infty$ and we conclude $\mathscr{P}(T) \supset \bigvee_{H(\xi) < \infty} \bigcap_n T^{-n} \hat{\xi}^-$.

We see from this theorem that $\mathscr{P}(T)$ contains the 'tails' of all finite entropy processes generated by countable finite-entropy partitions. In particular, if $\mathscr{P}(T) = \mathcal{N}$ then all tails are trivial. Hence if $\mathscr{P}(T) = \mathcal{N}$ and T has a finite-entropy generator ξ then T is a K-automorphism for $T^n \hat{\xi}^- \uparrow \mathscr{B}$ and $T^{-n} \hat{\xi}^- \downarrow \mathcal{N}$.

10. Theorem *If* $H(\xi) < \infty$ *and* $T^{-1} \mathscr{A} = \mathscr{A}$ *then*

$$H(\xi | \hat{\xi}^- \vee \mathscr{P}(T) \vee \mathscr{A}) = H(\xi | \hat{\xi}^- \vee \mathscr{A}).$$

Proof Choose $\hat{\eta} \subset \mathscr{P}(T)$ with $H(\eta) < \infty$; then

$$H(\xi \vee \eta | \hat{\xi}^- \vee \hat{\eta}^- \vee \mathscr{A}) = H(\eta | \hat{\eta}^- \vee \hat{\xi}_T \vee \mathscr{A}) + H(\xi | \hat{\xi}^- \vee \mathscr{A}),$$

i.e.

$$H(\xi | \hat{\xi}^- \vee \hat{\eta}^- \vee \mathscr{A}) = H(\xi | \hat{\xi}^- \vee \mathscr{A}).$$

Now let $H(\eta_n) < \infty$ and $\hat{\eta}_n \uparrow \mathscr{P}(T)$ to obtain

$$H(\xi | \hat{\xi}^- \vee \mathscr{P}(T) \vee \mathscr{A}) = H(\xi | \hat{\xi}^- \vee \mathscr{A}).$$

11. Theorem *If* $T^{-1} \mathscr{A} \subset \mathscr{A}$, $T^n \mathscr{A} \uparrow \mathscr{A}_\infty$, $T^{-n} \mathscr{A} \downarrow \mathscr{A}_{-\infty}$ *and* $\hat{\xi} \subset \mathscr{A}_\infty$, $H(\xi) < \infty$ *then*

$$H(\xi | \mathscr{P}(T) \vee \mathscr{A}_{-\infty}) = H(\xi | \mathscr{A}_{-\infty}).$$

Proof Assume $\hat{\xi} \subset \mathscr{A}$; then

$$H(\xi | \mathscr{P}(T) \vee \mathscr{A}_{-\infty}) \geq \lim_{p \to \infty} H(\xi | \hat{\xi}_{T^p}^- \vee \mathscr{A}_{-\infty} \vee \mathscr{P}(T))$$
$$= \lim_{p \to \infty} H(\xi | \hat{\xi}_{T^p}^- \vee \mathscr{A}_{-\infty})$$
$$= H(\xi | \mathscr{A}_{-\infty}).$$

(Here $\hat{\xi}_{T^p}^-$ is the σ-algebra generated by $\{T^{-pn}\hat{\xi}\}$, $n \geq 1$.) Hence

$$H(\xi \mid \mathscr{P}(T) \vee \mathscr{A}_{-\infty}) = H(\xi \mid \mathscr{A}_{-\infty}) \quad \text{for } \hat{\xi} \subset \mathscr{A}$$

and also for $\hat{\xi} \subset T^n \mathscr{A}$ any integer n.

To complete the proof we need: if $\hat{\xi} \subset \mathscr{A}_\infty$, $H(\xi) < \infty$, then for every $\varepsilon > 0$ there exists η with $\hat{\eta} \subset T^n \mathscr{A}$, $H(\eta) < \infty$ (for some n) such that $H(\xi \mid \hat{\eta}) + H(\eta \mid \hat{\xi}) < \varepsilon$. (See Exercise 12, Chapter 2.)

We now have, for ξ, η as above,

$$\begin{aligned}
H(\xi \mid \mathscr{A}_{-\infty}) &\leq H(\xi \vee \eta \mid \mathscr{A}_{-\infty}) \\
&\leq H(\eta \mid \mathscr{A}_{-\infty}) + H(\xi \mid \hat{\eta}) \\
&\leq H(\eta \mid \mathscr{A}_{-\infty} \vee \mathscr{P}(T)) + \varepsilon \\
&\leq H(\eta \vee \xi \mid \mathscr{A}_{-\infty} \vee \mathscr{P}(T)) + \varepsilon \\
&\leq H(\xi \mid \mathscr{A}_{-\infty} \vee \mathscr{P}(T)) + H(\eta \mid \xi) + \varepsilon \\
&\leq H(\xi \mid \mathscr{A}_{-\infty} \vee \mathscr{P}(T)) + 2\varepsilon.
\end{aligned}$$

Since $\varepsilon > 0$ is arbitrary we have

$$H(\xi \mid \mathscr{A}_{-\infty}) \leq H(\xi \mid \mathscr{A}_{-\infty} \vee \mathscr{P}(T)).$$

The reverse inequality is obvious.

Corollary *If $T^n \mathscr{A} \uparrow \mathscr{A}_\infty$ and $T^{-n} \mathscr{A} \downarrow \mathscr{N}$ then $\mathscr{A}_\infty$ is independent of $\mathscr{P}(T)$, i.e. $m(A \cap B) = m(A)\, m(B)$ for all $A \in \mathscr{A}_\infty$, $B \in \mathscr{P}(T)$.*

Proof We have $H(\xi \mid \mathscr{P}(T)) = H(\xi)$ for all $\hat{\xi} \subset \mathscr{A}_\infty$ $(H(\xi) < \infty)$. Therefore $\hat{\xi}$ is independent of $\mathscr{P}(T)$.

Corollary *If $T^n \mathscr{A} \uparrow \mathscr{B}$ then $\mathscr{P}(T) \subset \bigcap_n T^{-n} \mathscr{A} = \mathscr{A}_{-\infty}$.*

Proof For $H(\xi) < \infty$ we have

$$H(\xi \mid \mathscr{P}(T) \vee \mathscr{A}_{-\infty}) = H(\xi \mid \mathscr{A}_{-\infty}).$$

Therefore $\hat{\xi} \subset \mathscr{P}(T) \vee \mathscr{A}_{-\infty}$ implies $\hat{\xi} \subset \mathscr{A}_{-\infty}$, i.e. $\mathscr{P}(T) \subset \mathscr{A}_{-\infty}$.

Corollary *If T is a K-automorphism then T has completely positive entropy.*

12. Theorem (Rohlin and Sinai [1]) *There exists a sub-σ-algebra $\mathscr{A}$ such that $T^{-1} \mathscr{A} \subset \mathscr{A}$, $T^n \mathscr{A} \uparrow \mathscr{B}$, and $T^{-n} \mathscr{A} \downarrow \mathscr{P}(T)$. In particular (using the last corollary), an automorphism is a K-automorphism if and only if it has completely positive entropy.*

Proof Let $\hat{\xi}_n \uparrow \mathscr{B}$ where $H(\xi_n) < \infty$ and define, inductively, $\eta_p = \eta_{p-1} \vee T^{-n_p}\xi_p$ where the positive integers n_p are to be chosen (using Lemma 7) so that

$$H(\eta_1|\hat{\eta}_1^-) - H(\eta_1|\hat{\eta}_2^-) < \tfrac{1}{2}$$

and more generally

$$H(\eta_1|\hat{\eta}_{q-1}^-) - H(\eta_1|\hat{\eta}_q^-) < \frac{1}{2^{q-1}}$$

$$H(\eta_2|\hat{\eta}_{q-1}^-) - H(\eta_2|\hat{\eta}_q^-) < \frac{1}{2}\cdot\frac{1}{2^{q-2}}$$

$$\cdots\cdots\cdots$$

$$\cdots\cdots\cdots$$

$$H(\eta_{q-1}|\hat{\eta}_{q-1}^-) - H(\eta_{q-1}|\hat{\eta}_q^-) < \frac{1}{(q-1)2}$$

i.e. $H(\eta_p|\hat{\eta}_{q-1}^-) - H(\eta_p|\hat{\eta}_q^-) < (1/p)\cdot(1/2^{q-p})$, $\quad p = 1, \ldots, q-1$.
Now sum over $q = p+1,\ p+2, \ldots, n$, where p is fixed, to obtain

$$H(\eta_p|\hat{\eta}_p^-) - H(\eta_p|\hat{\eta}_n^-) < \frac{1}{p} \quad \text{for all } n > p.$$

Let $\eta_n \uparrow \mathscr{C}$ so that $\hat{\eta}_n^- \uparrow \mathscr{C}^-$, and let $\mathscr{A} = \mathscr{C}^-$. We have $H(\eta_p|\hat{\eta}_p^-)$ $- H(\eta_p|\mathscr{A}) \leq (1/p)$. Clearly $T^{-1}\mathscr{A} \subset \mathscr{A}$, $\quad T^n\mathscr{A} \uparrow \mathscr{B}$ and $\lim_{p\to\infty} (H(\eta_p|\hat{\eta}_p^-) - H(\eta_p|\mathscr{A})) = 0$. Let $\xi \subset \bigcap_n T^{-n}\mathscr{A}\,(H(\xi) < \infty)$, so that $\hat{\xi}_T \subset \bigcap_n T^{-n}\mathscr{A}$. Then

$$H(\xi|\hat{\xi}^-) = H(\eta_p \vee \xi|\hat{\eta}_p^- \vee \hat{\xi}^-) - H(\eta_p|\hat{\eta}_p^- \vee \hat{\xi}_T)$$

$$\leq H(\eta_p|\hat{\eta}_p^-) + H(\xi|\hat{\eta}_p^-) - H(\eta_p|\mathscr{A}).$$

As $p \to \infty$ we have

$$h(T, \xi) = H(\xi|\hat{\xi}^-) = 0.$$

Hence $\hat{\xi} \subset \bigcap_n T^{-n}\mathscr{A}$ implies $\hat{\xi} \subset \mathscr{P}(T)$, i.e. $\bigcap_n T^{-n}\mathscr{A} \subset \mathscr{P}(T)$. We clearly have $\bigcap_n T^{-n}\mathscr{A} \supset \mathscr{P}(T)$. The theorem is proved.

Corollary *If T is a K-automorphism then so are $T^p(p \neq 0)$. If T is a factor of a K-automorphism then T is a K-automorphism.*

13. Theorem *If $T^{-1}\mathscr{B}_n = \mathscr{B}_n \uparrow \mathscr{B}$ then $\mathscr{P}(T) \cap \mathscr{B}_n \uparrow \mathscr{P}(T)$.*

Proof Let $\hat{\xi} \subset \mathscr{P}(T)$, $H(\xi) < \infty$ and let $H(\eta) < \infty$; then

$$H(\xi \vee \eta | \hat{\eta}^-) = H(\xi | \hat{\eta}^-) + H(\eta | \hat{\xi} \vee \hat{\eta}^-)$$
$$= H(\eta | \hat{\eta}^-) + H(\xi | \hat{\eta} \vee \hat{\eta}^-).$$

But

$$H(\eta | \eta^-) \geq H(\eta | \hat{\xi} \vee \eta^-) \geq H(\eta | \hat{\eta}^- \vee \mathscr{P}(T)) = H(\eta | \hat{\eta}^-).$$

Therefore

$$H(\xi | \hat{\eta}^-) = H(\xi | \hat{\eta} \vee \hat{\eta}^-) = H(T^{-1}\xi | \hat{\eta}^-).$$

Now let $\hat{\eta}_k \subset \mathscr{B}_n, \hat{\eta}_k \uparrow \mathscr{A}_n$ where $T^l \mathscr{A}_n \uparrow \mathscr{B}_m$, $T^{-l}\mathscr{A}_n \downarrow \mathscr{B}_n \cap \mathscr{P}(T)$. We have $H(\xi | \mathscr{A}_n) = H(T^{-1}\xi | \mathscr{A}_n)$ or more generally $H(\xi | \mathscr{A}_n)$ $= H(T^p \xi | \mathscr{A}_n) = H(\xi | T^{-p} \mathscr{A}_n)$ whenever $\hat{\xi} \subset \mathscr{P}(T)$, $H(\xi) < \infty$. Taking limits as $p \to \infty$ and then as $p \to -\infty$ we have

$$H(\xi | \mathscr{B}_n) = H(\xi | \mathscr{B}_n \cap \mathscr{P}(T)).$$

Now let $n \to \infty$ to obtain $H(\xi | \vee_n(\mathscr{B}_n \cap \mathscr{P}(T))) = 0$, i.e.

$$\mathscr{P}(T) \subset \vee_n(\mathscr{B}_n \cap \mathscr{P}(T)) \subset \mathscr{P}(T).$$

14. Theorem* *If S, T are automorphisms then*

$$\mathscr{P}(S \times T) = \mathscr{P}(S) \times \mathscr{P}(T).$$

Proof Let $\mathscr{B}(S)$, $\mathscr{B}(T)$ be the measure σ-algebras for S, T and let $S^n \mathscr{A}(S) \uparrow \mathscr{B}(S)$, $S^{-n} \mathscr{A}(S) \downarrow \mathscr{P}(S)$, $T^n \mathscr{A}(T) \uparrow \mathscr{B}(T)$, $T^{-n} \mathscr{A}(T) \downarrow \mathscr{P}(T)$.

$$(S \times T)^n(\mathscr{A}(S) \times \mathscr{A}(T)) \uparrow \mathscr{B}(S) \times \mathscr{B}(T)$$

so that $\mathscr{P}(S \times T) \subset \bigcap_n[(S \times T)^{-n}(\mathscr{A}(S) \times \mathscr{A}(T))]$

$$\subset \bigcap_n[S^{-n}\mathscr{A}(S) \times \mathscr{B}(T)] = \mathscr{P}(S) \times \mathscr{B}(T),$$

and in the same way

$$\mathscr{P}(S \times T) \subset \mathscr{B}(S) \times \mathscr{P}(T),$$

i.e.

$$\mathscr{P}(S \times T) \subset \mathscr{P}(S) \times \mathscr{P}(T).$$

---

* Paul Shields informed me that this theorem was first proved by S. Pollit.

But

$$\mathcal{P}(S) \times \mathcal{N}(T) \subset \mathcal{P}(S \times T)$$
$$\mathcal{N}(S) \times \mathcal{P}(T) \subset \mathcal{P}(S \times T),$$

and therefore $\mathcal{P}(S) \times \mathcal{P}(T) \subset \mathcal{P}(S \times T)$.

In the proof of Theorem 14 we have used the not so obvious facts contained in:

Exercise 6 If $(X, \mathcal{A}, m), (Y, \mathcal{B}, p)$ are probability spaces and $\mathcal{A}_n \subset \mathcal{A}$ is a decreasing sequence of σ-algebras with $\mathcal{A}_n \downarrow \mathcal{A}_\infty$, then

$$\mathcal{A}_n \times \mathcal{B} \downarrow \mathcal{A}_\infty \times \mathcal{B}.$$

Exercise 7 With the hypotheses as above and $\mathcal{A}' \subset \mathcal{A}, \mathcal{B}' \subset \mathcal{B}$ (sub-σ-algebras), then $(\mathcal{A}' \times \mathcal{B}) \cap (\mathcal{A} \times \mathcal{B}') = \mathcal{A}' \times \mathcal{B}'$.

5 Zero entropy

15. Theorem *If T is an invertible measure-preserving transformation of the separable probability space* $(X, \mathcal{B}, m)$ *then either T has zero entropy or* U_T *has countable Lebesgue spectrum in the orthocomplement of* $L^2(X, \mathcal{P}(T), m)$.

Proof This is much the same as the proof of Theorem 6, Chapter 3. Let $T^n \mathcal{A} \uparrow \mathcal{B}$ and $T^{-n} \mathcal{A} \downarrow \mathcal{P}(T)$; then $\{U_T^n\}$ maps $V = L^2(X, \mathcal{A}, m) \ominus L^2(X, T^{-1} \mathcal{A}, m)$ onto a sequence of mutually orthogonal subspaces, and U_T has Lebesgue spectrum in the orthocomplement of $L^2(X, \bigcap_n T^{-n} \mathcal{A}, m)$ with multiplicity the same as the dimension of V, which is, by Lemma 5, Chapter 3, countably infinite.

Corollary *If T has a purely singular spectrum or a spectrum of finite multiplicity then* $h(T) = 0$.

Let $X = \prod_{n=-\infty}^{\infty} \mathbb{R}$ and let $\pi_n : X \to \mathbb{R}$ be the coordinate functions $\pi_n(x) = x_n, x = \{x_n\}$. X is given the σ-algebra $\mathcal{B}$ which is the smallest σ-algebra containing all sets $\pi_n^{-1} B$, with B a Borel subset of $\mathbb{R}$. As usual T will be the shift transformation, $T\{x_n\} = \{x_{n+1}\}$.

We consider any T-invariant probability m defined on $(X, \mathcal{B})$ such that $\int (\pi_0)^2 \, dm < \infty$. $(X, \mathcal{B}, m, T)$ is a stationary *real-valued stochastic process* with *finite second moments* (i.e. the random variables $\{\pi_n\}$ are square integrable). The class of stationary real-valued Gaussian processes is particularly important.

16. Theorem (Pinsker [1]; Newton [1]) *If T is a stationary real-valued process with finite second moments and* singular covariance, *then* $h(T) = 0$:

Proof The assumption here is that $\int \pi_n \pi_0 \, dm = \int \pi_0 (T^n) \pi_0 \, dm = \int_K \lambda^n \, d\tilde{\pi}_0$ defines a spectral measure $\tilde{\pi}_0$ on the circle K which is singular with respect to Lebesgue measure.

We note that π_0 is therefore orthogonal to $L^2(X, \mathscr{P}(T), m)^{\perp}$ since U_T restricted to this space has Lebesgue spectrum (unless it is trivial). Hence $\pi_0 \in L^2(X, \mathscr{P}(T), m)$ and, more generally, $\pi_n \in L^2(X, \mathscr{P}(T), m)$ for all n. Therefore $\{\pi_n\}$ are measurable with respect to $\mathscr{P}(T)$. But $\mathscr{B}$ is the smallest σ-algebra with respect to which $\{\pi_n\}$ are measurable. We conclude that $\mathscr{P}(T) = \mathscr{B}$, i.e. $h(T) = 0$.

Evidently an assumption such as finite multiplicity for the covariance would have led to the same conclusion.

We proceed now to some topological examples of zero entropy.

A homeomorphism T of a compact metric space X is said to be *distal* if for each $x, y \in X$, $x \neq y$, the closure of $\{(T^n x, T^n y)\}$ in $X \times X$ is disjoint from the diagonal $D = \{(z, z) : z \in X\}$.

17. Theorem (Parry [3]) *If T is a distal minimal homeomorphism of the compact metric space X which preserves the Borel probability m, then the entropy of T with respect to m is zero.*

Proof It suffices to construct a strong-generator ξ with finite entropy for then $h(T) = h(T, \xi) = H(\xi | \tilde{\xi}^-) = 0$ since $\tilde{\xi} \subset \tilde{\xi}^-$.

A partition ξ is a strong-generator if it is *strongly separating* in the sense that $T^n x$, $T^n y \in A_{i_n} \in \xi$ for $n = 0, 1, 2, \dots$ implies $x = y$ (cf. Exercise 8, this Chapter).

Let $x_0 \in X$ and let $S_n = \{x \in X : d(x, x_0) < r_n\}$ where r_n are chosen to be decreasing and so that $m(S_n) < e^{-n}$, $n = 1, 2, \ldots$, and therefore

$$- m(S_n - S_{n+1}) \log m(S_n - S_{n+1}) < ne^{-n}.$$

(This can be done except in the trivial case that $m\{x_0\} > 0$, i.e. except when X is a finite space, which can be dealt with easily.)

Now let $\xi = (A_1, A_2, \ldots, A_\infty)$ $A_n = S_n - S_{n+1}$, $A_\infty = \{x_0\}$. Evidently $H(\xi) < \sum_{n=1}^{\infty} e^{-n} n < \infty$. ξ is strongly separating for if $T^n x$, $T^n y \in A_{i_n}$, $n = 0, 1, 2, \ldots$, then $T^n x$, $T^n y \in S_{i_n}$, $n = 0, 1, 2, \ldots$. But the sequence $\{i_n\}$ is unbounded since T is minimal and therefore $\{T^n x\}$ comes arbitrarily close to x_0. Hence the pair $T^n x$, $T^n y$ belongs to arbitrarily small neighbourhoods of x_0 as n runs through the positive integers. We conclude that (x_0, x_0) is in the closure of $\{(T^n x, T^n y): n = 0, 1, \ldots \}$. By the distality of T, $x = y$ and the proof is complete.

Exercise 8 Show that a partition which is strongly separating is a strong-generator.

Exercise 9 Show that $T(x_1, \ldots, x_k) = (\alpha_1 + x_1, \ldots, \alpha_k + x_k)$ (mod 1) is an isometry of the k-torus and is therefore distal. Deduce that $h(T) = 0$ when $\alpha_1, \ldots, \alpha_k$, 1 are rationally independent. (This latter assumption is actually not necessary.)

Exercise 10 Show that $T(x, y) = (\alpha + x, x + y)$ (mod 1) is a distal homeomorphism which is not an isometry (for any compatible metric). If α is irrational (again this is unnecessary) deduce that $h(T) = 0$.

5

Some examples

1 Flows and changes in velocity

Let $\{T_t: -\infty < t < \infty\}$ be a one-parameter group of measure-preserving transformations of the probability space $(X, \mathscr{B}, m)$ onto itself, where the map $X \times \mathbb{R} \to X$ $((x, t) \to T_t x)$ is measurable. The *flow* $\{T_t\}$ is said to be *ergodic* if $f \circ T_t = f$ a.e. for each t (the null set may depend on t) implies f is constant a.e. If $f \circ T_t = \exp(2\pi i b t) \cdot f$ a.e. (*b* a constant) for all t, where $0 \neq f \in L^2(X)$, then b is called an *eigenfrequency* and f is called an *eigenfunction*. If every eigenfunction is constant (a.e.) the flow is said to be *weak-mixing*. A flow $\{T_t\}$ is said to be *strong-mixing* if $\int (f \circ T_t)\bar{g}\,\mathrm{d}m \to \int f\,\mathrm{d}m \int \bar{g}\,\mathrm{d}m$ as $t \to \infty$ for all f, $g \in L^2(X)$. Alternative, equivalent definitions of ergodicity, weak-mixing, and strong-mixing can be given in a manner analogous to those formulated for a single transformation in Chapter 3.

If $\{T_t\}$ is a flow we can obtain from it a new flow $\{S_t\}$ with exactly the same orbits traversed in the same direction but at different speeds. In general, $\{S_t\}$ will not preserve the measure m associated with $\{T_t\}$. It will preserve some equivalent measure, however. We proceed as follows. Let $k: X \times \mathbb{R} \to \mathbb{R}$ be positive, measurable and satisfy

$$k(x, s + t) = k(T_t x, s) + k(x, t).$$

Such a function is called an (additive) positive *cocycle*. More precisely we let $k(x, t) = \int_0^t k(T_u x)\,\mathrm{d}u$ (we use the same symbol and hope this will not cause confusion) where $k(x)$ is positive, measurable and bounded away from zero and infinity. We assume also that $\int k\,\mathrm{d}m = 1$. Now let $h(x, t)$ be defined by

$$t = \int_0^{h(x, t)} k(T_u x)\,\mathrm{d}u = \int_0^{h(x, t)} \frac{1}{h(T_u x)}\,\mathrm{d}u$$

where $h(x) = 1/k(x)$. (Again we use h for two functions.)
Define $S_t(x) = T_{h(x, t)}(x)$ and note that S_t is a transformation of $(X, \mathscr{B})$ for each $t \in \mathbb{R}$. Moreover $S_{u+v} = S_u \circ S_v$ since $T_{h(x, u+v)}(x)$

$= T_{h(S_v x, u)}(T_{h(x, v)}(x))$ because $h(x, u + v) = h(S_v x, u) + h(x, v)$, i.e. h is an additive cocycle for $\{S_t\}$. In fact h and k are inverse to each other in the sense that $h(x, k(x, t)) = t$ a.e. for each t. This follows from

$$k(x, t) = \int_0^{h(x, k(x, t))} k(T_u x)\, du = \int_0^t k(T_u x)\, dx.$$

The fact that h is a cocycle for $\{S_t\}$ now follows from the fact that k is a cocycle for $\{T_t\}$. In short, $\{S_t\}$ is a one-parameter group of measurable transformations of $(X, \mathscr{B})$. It is not difficult to see that the map $X \times \mathbb{R} \to X$ $((x, t) \to S_t x)$ is measurable. Finally the transformations S_t preserve the measure $k\, dm$, when T_t is assumed to be ergodic. In fact it suffices to note that for $f \in L^1(X, m) = L^1(X, k\, dm)$

$$\frac{1}{U}\int_0^U f(S_t x)\, dt = \frac{1}{U}\int_0^U f(T_{h(x, t)}(x))\, dt$$

$$= \frac{1}{U}\int_0^{h(x, U)} f(T_s x)k(T_s x)\, ds \quad (s = h(x, t),\ t = k(x, s))$$

$$= \frac{1}{k(x, V)}\int_0^V f(T_s x)\, k(T_s x)\, ds \quad (U = k(x, V))$$

$$= \left(\int_0^V f(T_s x)\, k(T_s x)\, ds\right)\bigg/\left(\int_0^V k(T_s x)\, ds\right).$$

Therefore taking limits and applying the Birkhoff ergodic theorem for flows we have

$$\lim_{U \to \infty} \frac{1}{U}\int_0^U f(S_t x)\, dt = \frac{\int f k\, dm}{\int k\, dm} = \int f k\, dm$$

$$= \lim_{U \to \infty} \frac{1}{U}\int_0^U f(S_{t + t_0} x)\, dt = \int (f \circ S_{t_0}) k\, dm.$$

It is clear that $\{S_t\}$ is ergodic since it has the same orbits as $\{T_t\}$. Summarising the above we have:

1. Theorem *For h, k, defined as above, with $\{T_t\}$ ergodic, the flow $\{S_t\}$ $(S_t x = T_{h(x, t)} x)$ preserves the measure $k\, dm$ and is ergodic.*

2 Winding numbers

Let X be a compact metric space and let m be a probability defined on the Borel σ-algebra $\mathscr{B}$ of X. A one-parameter group of homeomorphisms $\{T_t\}$ is said to be a *continuous flow* (with respect to m) if the map $X \times \mathbb{R} \to X$ $((x, t) \to T_t x)$ is continuous and if each T_t is measure-preserving.

Schwartzman [1] and [2], generalizing an idea due to Poincaré, has introduced an important invariant for continuous flows. The invariant is a canonical 1-dimensional real homology class i.e. a canonical homomorphism W_m of the first (Čech) cohomology group $H^1(X, \mathbb{Z})$ into $\mathbb{R}$. The image $W_m(\{T_t\}) = W_m H^1(X, \mathbb{Z})$ is called the *winding numbers group* and each element of $W(\{T_t\})$ is called *a winding number*.

It is known that $H^1(X, \mathbb{Z}) \simeq \mathrm{Br}(X)$, the Bruschlinsky group consisting of the group of continuous maps of X into $K = \{|z| = 1\}$, factored by the group of maps which are homotopic to a constant map. This group is countable (due to the fact that X is separable).

Let $f: X \to K$ be continuous, then there exists $g \in \tilde{f}$ (the coset or class of functions homotopic to f) such that g is continuously differentiable with respect to the flow $\{T_t\}$, i.e.

$$g'(x) = \lim_{t \to 0} \left[\frac{g(T_t x) - g(x)}{t} \right]$$

exists for all $x \in X$ and $g'(x)$ is continuous. If g is continuously differentiable and homotopic to a constant, then $g(x) = \exp 2\pi i r(x)$ $r: X \to \mathbb{R}$, and r is continuously differentiable. Hence

$$\frac{1}{2\pi i} \int \frac{g'(x)}{g(x)} \, \mathrm{d}m = \int r'(x) \, \mathrm{d}m = \int \lim_{t \to 0} \left[\frac{r(T_t x) - r(x)}{t} \right] \mathrm{d}m = 0,$$

since $\{T_t\}$ is measure-preserving. This allows us to define unambiguously for each $\tilde{f} \in \mathrm{Br}(X)$

$$W_m(\tilde{f}) = W_m(g) = \frac{1}{2\pi i} \int \frac{g'(x)}{g(x)} \, \mathrm{d}m,$$

when $g \in \tilde{f}$ is continuously differentiable. In fact if g, h are continuously differentiable and $g, h \in \tilde{f}$ then $g = h \cdot k$ where k is continuously differentiable and homotopic to a constant. Hence

$$\frac{1}{2\pi i} \frac{g'}{g} = \frac{1}{2\pi i} \frac{h'}{h} + \frac{1}{2\pi i} \frac{k'}{k}$$

and

$$W_m(g) = W_m(h) + W_m(k) = W_m(h).$$

($g'/2\pi i g$ measures the rate of change of angle in K with varying time – hence, this is a *real*-valued function.)

Clearly for all $\tilde{f}, \tilde{g} \in \mathrm{Br}(X)$, $W_m(\tilde{f} \cdot \tilde{g}) = W_m(\tilde{f}) + W_m(\tilde{g})$. In other words, W_m is a canonical homomorphism of $H^1(X, \mathbb{Z}) \simeq \mathrm{Br}(X)$ into $\mathbb{R}$ and thus represents a 1-dimensional homology class.

2. Theorem *If T_t is a continuous flow on the compact metric probability space $(X, \mathscr{B}, m)$ and if k is a positive continuous function on X which is bounded away from zero with $\int k \, dm = 1$, then W_m defined with respect to T_t is identical to W_p defined with respect to S_t (the flow obtained from T_t by a change of velocity effected by k) where $(dp/dm) = k$.*

Proof Let $g \in \tilde{f} \in \mathrm{Br}(X)$ be continuously differentiable then its derivative g^*, with respect to S_t, is given by

$$g^*(x) = \lim_{t \to 0} \frac{g(S_t x) - g(x)}{t}$$

$$= \lim_{t \to 0} \frac{g(T_{h(x, t)}(x) - g(x))}{t}$$

$$= \lim_{t \to 0} g'(x) \frac{h(x, t)}{t}$$

$$= g'(x)/k(x)$$

where g' is the derivative of g with respect to T_t. Hence

$$W_p(g) = \frac{1}{2\pi i} \int \frac{g^*}{g} k \, dm = \frac{1}{2\pi i} \int \frac{g'(x)}{g(x)} \, dm$$

$$= W_m(g).$$

We have remarked that ergodicity remains invariant under a change of velocity. For continuous flows it is clear that unique ergodicity is also invariant under a change of velocity, since the correspondence $dm \to k \, dm$ transforms ergodic measures for T_t into ergodic measures for S_t. Minimality is clearly another property which remains invariant. We proceed now to investigate discrete spectrum and show that it is possible (in general) to change velocity so that an ergodic flow becomes weak-mixing. This is a result of Chacon [1] in answer to a question posed by Hopf. The method we adopt, however, has some advantages, in that the general case can have its velocity changed 'continuously' to abolish eigenvalues. The first result in this more topological direction was achieved by Humphreys [1]. Related to this problem is an example due to Kolmogorov, of a velocity change which has discrete spectrum yet has no non-constant *continuous* eigenfunctions.

We proceed to Chacon's problem which we solve with the aid of a theorem of Helson and Kahane.

3 Abolishing eigenvalues

3. Lemma *If T_t is an ergodic measure-preserving flow on the separable probability space $(X, \mathcal{B}, m)$ then there exists $t_n \to \infty$ with*

$$\int f(T_{t_n} x) \overline{f(x)} \, dm \to 0 \quad \text{for all} \quad f \in V_c$$

and

$$\int f(T_{t_n} x) \, \overline{f(x)} \, dm \to \int |f|^2 \, dm \quad \text{for all } f \in V_d$$

where $L^2(X) = V_d \oplus V_c$ is the orthogonal decomposition of $L^2(X)$ respecting discrete and continuous spectra.

Proof We remark without proof (cf. Pugh and Shub [1]) that it is easy to show that for all but at most a countable number of t, the individual transformations T_t are ergodic. Let T_{t_0} be ergodic, and let $Uf = f \circ T_{t_0}$.

$V_c \oplus V_d$ also respects the continuous and discrete spectra of U. Let $\{f_n\}$ be a countable orthonormal basis of V_d consisting of eigenfunctions, and let $\{g_n\}$ be a countable sequence dense in V_c. For each k, $N_k = \{n > 0 : |\langle U^n f_i, f_i \rangle - 1| < (1/k), i = 1, \ldots, k\}$ is a relatively dense sequence of positive integers (i.e. for some $l > 0$, N_k intersects *every* · interval of positive integers of length l).

For some subset J_k of density zero of the positive integers

$$\lim_{\substack{n \to \infty \\ n \notin J_k}} \langle U^n g_k, g_k \rangle = 0.$$

Therefore $M_k = N_k - J_1 \cup \cdots \cup J_k$ is infinite for each k, and $M_1 \supset M_2 \supset \cdots$ Let $t_1 < t_2 < \cdots$, where $t_i \in M_i$. Then

$$\lim_{n \to \infty} \langle U^{t_n} f_i, f_i \rangle = 1 \quad \text{and} \quad \lim_{n \to \infty} \langle U^{t_n} g_i, g_i \rangle = 0$$

for $i = 1, 2, \ldots$. A standard approximation argument completes the proof.

4. Lemma (Helson and Kahane [1]) *Let Γ be a subgroup of the group of eigenfrequencies of the ergodic flow $\{T_t\}$. Suppose Γ is dense in the real line and let $t_n \to \infty$ be a sequence such that*

$$\int f(T_{t_n} x) \overline{f(x)} \, dm \to \int |f|^2 \, dm = 1$$

for all unit modulus eigenfunctions f. Then there exists $0 < \lambda_n \in \Gamma$ with $\sum_{n=1}^{\infty} \lambda_n < \infty$ and a subsequence $\{u_n\}$ of $\{t_n\}$ with

$$\left| \sum_{j=-\infty}^{\infty} f_j(x)(1 - \exp(2\pi i u_n \lambda_j)) - 4 \mathcal{R} f_n(x) \right|$$

converging uniformly to zero, where $\lambda_{-j} = -\lambda_j$, $\lambda_0 = 0$ and $f_n(T_t x) = \exp(2\pi i \lambda_n t) f_n(x)$.

Proof Suppose $\lambda_1, \ldots, \lambda_n$, $u_1, \ldots, u_n$ have been chosen so that $u_1 < \cdots < u_n$, $\{\lambda_i\} \subset \Gamma$, $0 < \lambda_i < (1/2^i)$ $\{u_i\} \subset \{t_i : i = 1, 2, \ldots\}$, and for all $m = 1, \ldots, n$

$$\langle\langle u_m \lambda_i \rangle\rangle < (1/2^{m+i}) \quad i \neq m,$$

$$\langle\langle u_m \lambda_m + 1/2 \rangle\rangle < (1/2^m),$$

where $\langle\langle \ \rangle\rangle$ means the distance from the nearest integer. (This is easy to arrange for $n = 1$.) We proceed to define λ_{n+1}, u_{n+1}. Since, as $N \to \infty$,

$$\int f_i \circ T_{t_N} \bar{f}_i \, dm = \exp(i\lambda_i t_N 2\pi) \int |f_i|^2 \, dm \to \int |f_i|^2 \, dm$$

we have, for all sufficiently large N, $\langle\langle \lambda_i t_N \rangle\rangle$ $(i = 1, \ldots, n)$ as small as we please. The interval $(0, 1/(u_n 4^{n+1}))$ is contained in $\bar{\Gamma}$ so that, for even larger N, there exists t_N with $\langle\langle \lambda_i t_N \rangle\rangle < 1/2^{i+n+1}$, $i \leq n$, and $\langle\langle \lambda_{n+1} t_N + 1/2 \rangle\rangle < 1/2^{n+1}$ for some $\lambda_{n+1} \in (0, 1/(u_n 4^{n+1}))$. We take $u_{n+1} = t_N$. Clearly

$$\langle\langle \lambda_{n+1} u_{n+1} + 1/2 \rangle\rangle < 1/2^{n+1},$$

$$0 < \lambda_{n+1} u_i < 1/4^{n+1} < 1/2^{n+1+i}, \quad 1 \leq i < n+1,$$

and

$$\langle\langle \lambda_i u_{n+1} \rangle\rangle < 1/2^{n+1+i}, \quad 1 \leq i < n+1.$$

By induction we have two sequences $\{\lambda_i\} \subset \Gamma$, $0 < \lambda_i < 1/2^i$, $\{u_i\} \subset \{t_i\}$ such that $\langle\langle u_m \lambda_n \rangle\rangle < 1/2^{m+n} (m \neq n)$, $\langle\langle u_n \lambda_n + 1/2 \rangle\rangle < 1/2^n$. We define $\lambda_{-j} = -\lambda_j$. Evidently

$$\left| \sum_{j=-\infty}^{\infty} f_j(x) [1 - \exp(2\pi i \lambda_j u_n)] - 4 \mathcal{R} f_n \right|$$

$$< |2\mathcal{R}(f_n(x)[1 - \exp(2\pi i \lambda_n u_n)]) - 4\mathcal{R}(f_n)|$$

$$+ \sum_{|j| \neq n} |1 - \exp(2\pi i \lambda_j u_n)|$$

$$< 2|1 + \exp(2\pi i \lambda_n u_n)| + 2 \sum_{\substack{j > 0 \\ j \neq n}} |1 - \exp(2\pi i \lambda_j u_n)|$$

$$< (2/2^n) + 2 \sum_{j=1}^{\infty} (1/2^{n+j}) \to 0 \quad \text{as} \quad n \to \infty.$$

5. Theorem (Chacon [1]; Parry [4]) *Let $\{T_t\}$ be an ergodic flow on the separable probability space $(X, \mathcal{B}, m)$ with an eigenfrequency group which is dense in $\mathbb{R}$. (In other words the eigenfrequency group is not cyclic.) Then there is a change of velocity so that the new flow is weak-mixing. (Chacon did not require the acyclicity condition.)*

Proof Let Γ be a subgroup of the eigenfrequency group which is dense in $\mathbb{R}$. Let $t_n \to \infty$ be a sequence satisfying Lemma 3, i.e.

$$\int (f \circ T^{t_n}) f \, dm \to \int |f|^2 \, dm \quad \text{for all eigenfunctions}$$

and

$$\int (f \circ T^{t_n}) f \, dm \to 0 \quad \text{for all } f \in V_c.$$

Choose $\{\lambda_n\} \subset \Gamma$ and $\{u_n\} \subset \{t_n\}$ in accordance with Lemma 4. Define $k_0(x) = \sum_{j=-\infty}^{\infty} 2\pi i \lambda_j f_j(x)$; since $\sum_j |\lambda_j| < \infty$ and $\lambda_{-j} \overline{f_{-j}(x)} = -\lambda_j \overline{f_j(x)}$ we see that k_0 is real and bounded. Therefore for some a, $k(x) = a + k_0(x)$ is positive and bounded away from zero and infinity. We propose to change the velocity of $\{T_t\}$ with respect to k. In other words let

$$k(x, t) = \int_0^t k(T_s x) \, ds$$

$$= at + \sum_{j=-\infty}^{\infty} 2\pi i \lambda_j \int_0^t \exp(2\pi i \lambda_j s) \cdot f_j(x) \, ds$$

$$= at + \sum_{j=-\infty}^{\infty} 2\pi i \lambda_j f_j(x) \left[\frac{\exp(2\pi i \lambda_j t) - 1}{2\pi i \lambda_j} \right]$$

$$= at + \sum_{j=-\infty}^{\infty} f_j(x) [\exp(2\pi i \lambda_j t) - 1],$$

and define $h(x, t)$ in accordance with §1. Let $S_t(x) = T_{h(x, t)}(x)$; then we show that there are no non-constant eigenfunctions for S_t. Suppose $F(S_t x) = \exp(ibt) F(x)$, $F \in L^2(X, \mathcal{B}, k \, dm)$; we can assume $|F(x)| \equiv 1$, since $|F|$ is constant. Evidently

$$F(T_t x) = \exp[ibk(x, t)] F(x). \tag{5.1}$$

Let $\mathcal{A}$ be the σ-algebra generated by all eigenfunctions of $\{T_t\}$, i.e. the smallest $\{T_t\}$-invariant σ-algebra with respect to which all eigenfunctions are measurable. Then

$$E(F|\mathcal{A}) \circ T_t = \exp[ibk(x, t)] E(F|\mathcal{A}) \tag{5.2}$$

since $k(x, t)$ is $\mathcal{A}$-measurable. Comparing this equation with (5.1), and using ergodicity, we see that either F is $\mathcal{A}$-measurable or $E(F|\mathcal{A}) = 0$. In the first case we have

$$\int F \circ T_{t_n} \bar{F} \, dm \to \int |F|^2 \, dm = 1,$$

and in the second case we have

$$\int F \circ T_{t_n} \bar{F} \, dm \to 0.$$

In any case we have

$$\int \exp[ibk(x, u_n)] \, dm \to 0 \quad \text{or} \quad 1,$$

and therefore, applying Lemma 4,

$$\left| \int \exp[-4ib\mathcal{R}(f_n)] \, dm \right| \to 0 \quad \text{or} \quad 1.$$

Expanding the exponential and using the fact that $\{f_n\}$ is an orthonormal sequence, we obtain

$$\sum_{n=0}^{\infty} \frac{(2ib)^{2n}}{(n!)^2} = J_0(4b) = 0 \quad \text{or} \quad 1,$$

where J_0 is the zero-order Bessel function. But the above argument also applies to the eigenfrequencies $\{nb : n = 0, \pm 1, \ldots\}$. Hence $|J_0(4nb)| = 0$ or 1 for all integers n. However, $|J_0(4nb)| = 1$ (for some $n \neq 0$) implies $b = 0$ and the alternative is impossible since it contradicts the asymptotic formula for the zeros of the Bessel function (cf. Watson [1]).

6. Theorem *Let* $\{T_t\}$ *be an ergodic continuous flow on a compact metric space with Borel probability* m. *If there is a non-cyclic group of* *eigenfrequencies corresponding to continuous eigenfunctions, then the velocity can be changed by a strictly positive continuous function k to yield a weak-mixing flow.*

Proof It suffices to note that if the eigenfunctions f_n, corresponding to eigenfrequencies λ_n, in Theorem 5, are continuous, then the function $k_0(x) = \sum_{n=-\infty}^{\infty} 2\pi i \lambda_n f_n$ is continuous.

4 Discrete spectrum with 'topological weak-mixing'

We devote this section to an example of a flow on the 2-torus which is ergodic with discrete spectrum but which has no non-constant continuous eigenfunctions. The example, which is due to Kolmogorov, is obtained from an irrational flow by a change of velocity.

Let
$$T_t(x, y) = (e^{2\pi i \alpha t} x, e^{2\pi i t} y),$$

$x, y \in K$, where the irrational α has the form $\alpha = \sum_{l=1}^{\infty} 1/2^{\nu_l}$ and $\nu_{l+1} - \nu_l = \theta_l$ a positive integer. Let $k_0 : K \times K \to \mathbb{R}$ have the Fourier series

$$k_0(x, y) = \sum_{l=1}^{\infty} \frac{(m_l \alpha - n_l)}{l} (x^{m_l} y^{-n_l} + x^{-m_l} y^{n_l})$$

where $m_l = 2^{\nu_l}$, $n_l = [2^{\nu_l} \alpha]$ so that

$$m_l \alpha - n_1 = 2^{\nu_l} \alpha - [2^{\nu_l} \alpha] = \sum_{j=l+1}^{\infty} (2^{\nu_l}/2^{\nu_j}) \le (2 \cdot 2^{\nu_l})/2^{\nu_{l+1}} = 2/2^{\theta_l}.$$

If $\theta_l \to \infty$ we see that the Fourier coefficients are absolutely summable and therefore $k_0(x, y)$ is continuous.

We consider the function equation

$$f\, T_t(x, y) - f(x, y) = \int_0^t k_0 T_s(x, y)\, ds \qquad (5.3)$$

and show that whereas it has an $L^2(K \times K)$ solution f, it has no continuous solution f.

In fact writing

$$f(x, y) = 2\mathcal{R} \sum_{l=1}^{\infty} b_l x^{m_l} y^{-n_l},$$

(5.3) becomes:

$$2\mathscr{R} \sum_{l=1}^{\infty} b_l \{\exp[2\pi i(m_l\alpha - n_l)t] - 1\} x^{m_l} y^{-n_l}$$

$$= 2\mathscr{R} \sum_{l=1}^{\infty} \frac{(m_l\alpha - n_l)}{l} \left\{ \frac{\exp[2\pi i(m_l\alpha - n_l)t] - 1}{2\pi i(m_l\alpha - n_l)} \right\} x^{m_l} y^{-n_l}.$$

Putting $b_l = 1/(2\pi il)$ we have an $L^2(K \times K)$ solution of (5.3). It is not difficult to see from (5.3) that any other solution must equal

$$f(x, y) = 2\mathscr{R} \sum_{l=1}^{\infty} \frac{1}{2\pi il} x^{m_l} y^{-n_l} \quad \text{a.e.}$$

and therefore cannot be continuous. (Abel's test (cf. Zygmund [1]) will verify that the series is not the Fourier series of a continuous function.) Now let $k(x, y) = 1 + \varepsilon k_0(x, y)$ where ε is small enough to ensure that the continuous function $k(x, y)$ is strictly positive. We change the velocity of T_t (in accordance with §1) with respect to k to obtain an ergodic flow S_t preserving the measure $k\,dx\,dy$. (Note that $\int k\,dx\,dy = 1$.)

7. Theorem (Kolmogorov [4]) *The flow S_t has discrete spectrum but no non-constant continuous eigenfunctions.*

Proof For each pair of integers (m, n), $F(x, y) = x^m y^n \exp[2\pi i\delta f(x, y)]$ is an eigenfunction corresponding to the eigenfrequency $m\alpha + n$, where $\delta = (m\alpha + n)/\varepsilon$, since $F \circ S_t = \exp[2\pi it(m\alpha + n)]F$ is equivalent to $F \circ T_t = \exp[2\pi i(m\alpha + n)k(x, y, t)]F$, where

$$k(x, y, t) = \int_0^t k(T_s(x, y))\,ds$$

$$= t + \varepsilon \int_0^t k_0(T_s(x, y))\,ds.$$

Substitution, using (5.3), easily verifies $F \circ T_t = \exp[2\pi i(m\alpha + n)k(x, y, t)]F$. Clearly the eigenfunctions $(x, y) \to x^m y^n \exp[2\pi if(x, y)\delta]$ form a basis for $L^2(K \times K)$. Hence S_t has discrete spectrum. We show now that S_t has no non-constant continuous eigenfunctions. Suppose $F \circ T_t = \exp[2\pi i\gamma k(x, y, t)]F$, $\gamma \neq 0$, where F is continuous. We may suppose $|F| = 1$. Then F is homotopic to one of the functions $(x, y) \to x^m y^n$, i.e. $F(x, y) = x^m y^n \exp[2\pi i r(x, y)]$ where $r : K \times K \to \mathbb{R}$ is continuous. Hence

$$x^m y^n \exp[2\pi i (m\alpha + n)t] \exp[2\pi i r(T_t(x, y))]$$

$$= \exp[2\pi i \gamma k(x, y, t)] x^m y^n \exp[2\pi i r(x, y)].$$

We conclude that

$$(m\alpha + n)t + r(T_t(x, y)) = r(x, y) + \gamma k(x, y, t)$$

$$= r(x, y) + \gamma t + \gamma \varepsilon k_0(x, y, t).$$

Integration shows that $m\alpha + n = \gamma$ and therefore

$$r(T_t(x, y)) - r(x, y) = \gamma \varepsilon k_0(x, y, t)$$

contradicting the fact that (5.3) has no continuous solution.

5 Minimality without unique ergodicity

A similar analysis to Kolmogorov's was carried out by Furstenberg [1] who produced an example of a homomorphism of the 2-dimensional torus which is minimal but not uniquely ergodic.

If S is a minimal homeomorphism of the compact metric space X and φ is a continuous map of X to K then $T : X \times K \to X \times K$ given by $T(x, y) = (Sx, \varphi(x) y)$ is minimal if $F(Sx) = \varphi(x)^l F(x)$ $(l \neq 0)$ has no continuous solution F with $|F| = 1$. (The converse is also true but we shall not need this fact.) The proof, which we sketch, proceeds as follows: if T is not minimal then there is a proper minimal subset M of $X \times K$. Writing $g(x, y) = (x, gy)$ it is easy to see that $gM = M$ or $gM \cap M = \varnothing$ for all $g \in K$, and $\bigcup_{g \in K} gM = X \times K$. Let H be the proper closed subgroup of K consisting of those g for which $gM = M$. H is necessarily finite so that there exists l such that $g^l = 1$ for all $g \in H$. Now define $F(x) = g^l$ if $(x, g) \in M$. (This is well defined.) Since $(x, g) \in M$ implies $(Sx, \varphi(x)g) \in M$ we have $F(Sx) = \varphi(x)^l g^l = \varphi(x)^l F(x)$.

If S is an ergodic measure-preserving transformation of the probability space $(X, \mathscr{B}, m)$ and $\varphi : X \to K$ is measurable, then $T : X \times K \to X \times K$ given by $T(x, y) = (Sx, \varphi(x) y)$ is *not* ergodic with respect to $m \times l$ (l Lebesgue measure), if

$$F(Sx) = \varphi(x)^l F(x) \quad \text{a.e.} \quad (l \neq 0)$$

has a measurable solution F with $|F| = 1$. (Again the converse is true, but we shall not need that fact.) The proof is easy. In fact (x, y) $\to F(x)y^{-l}$ is T-invariant, since $(Sx, \varphi(x)y) \to F(Sx)\varphi(x)^{-l}y^{-l}$ $= F(x)y^{-l}$, and this function is not constant a.e.

We proceed now to Furstenberg's construction. Let $T(x, y)$ $= (e^{2\pi i\alpha}x, e^{2\pi i r(x)}y)$ where α is irrational and where $r: K \to \mathbb{R}$ is continuous. By the preceding remarks T is minimal if the equation

$$F(e^{2\pi i\alpha}x) = e^{2\pi i l r(x)}F(x) \quad (l \neq 0) \tag{5.4}$$

has no continuous solution F with $|F| = 1$. On the other hand, T is not ergodic (and therefore not uniquely ergodic) if (5.4) has a measurable solution (a.e.) F with $|F| = 1$ for some $l \neq 0$.

We see then, by putting $F(x) = e^{2\pi i f(x)}$ that T will be minimal and not uniquely ergodic if the equation

$$f(e^{2\pi i\alpha}x) - f(x) = r(x)$$

has an $L^2(K)$ solution but no continuous solution $f: K \to \mathbb{R}$. (Every continuous map from K to K is homotopic to one of the functions, $x \to x^n$.)

By a suitable choice of α and r this problem is solved in much the same way as in Kolmogorov's problem, so that we have:

8. Theorem (Furstenberg [1]) *There exists a minimal non-uniquely ergodic homeomorphism of the 2-dimensional torus.*

6 Lebesgue spectrum from discrete spectrum

In this section we produce an example of an ergodic measure-preserving transformation with discrete spectrum for which there is a simple modification – a so-called $\mathbb{Z}_2$ extension – which has, in addition to the discrete spectrum, countable Lebesgue spectrum.

Consider the sequence of finite cyclic groups $\{\mathbb{Z}_{\theta_n}\}$ where the positive integers θ_n are co-prime. Define $X = \prod_{n=1}^{\infty} \mathbb{Z}_{\theta_n}$ and let m be the direct product measure obtained from the measures m_n which assign measure $1/\theta_n$ to each of the points of $\mathbb{Z}_{\theta_n}$. Let $1 = (1, 1, \dots) \in X$ be the sequence whose coordinates 1 are generators of the groups $\mathbb{Z}_{\theta_n}$. X is a compact abelian group so we can define $Tx = x + 1, x = \{x_n\}$. It is not difficult to prove that T is ergodic (due to the co-prime condition) and has discrete spectrum. In fact defining the primitive character of $\mathbb{Z}_{\theta_n}$ by $\omega_n(x_n) = \exp(2\pi i x_n/\theta_n)$, each of the characters of X defined by $x \to \omega_n(x_n)$ is an eigenfunction with eigenvalue $\exp(2\pi i/\theta_n)$. Arbitrary finite products of these characters (again characters) are also eigenfunctions and form an orthonormal basis for $L^2(X)$. Ergodicity can be proved by Fourier analysis with respect to this orthonormal basis.

More specifically let $\theta_n = p_n^2$ where $p_1, p_2, \ldots$ is the sequence of primes, and represent $\mathbb{Z}_{p_n^2}$ by $(0, 1, \ldots, p_n^2 - 1)$ with mod p_n^2 addition. Let $A_n \subset X$ be defined by $A_n = \{x \in X : x_n = 0\}$ so that $m(A_n) = 1/p_n^2$ and $\sum_{n=1}^{\infty} \chi_{A_n}(x) = \chi(x)$ is summable.

9. Theorem (Helson and Parry [1]) *The transformation of $X \times \mathbb{Z}_2$ $(\mathbb{Z}_2 = \{\pm 1\})$ to itself defined by $S(x, y) = (Tx, e^{\pi i \chi(x)} y)$ (is ergodic and) has countable Lebesgue spectrum in the orthocomplement of $L^2(X) \subset L^2(X \times \mathbb{Z}_2)$.*

Proof It will suffice to show that

$$\int \exp[\pi i(\chi + \cdots + \chi \circ T^{n-1})] \, dm = \int_K z^n \, d\mu$$

where μ is absolutely continuous with respect to Lebesgue measure on K. For then we can argue as follows: divide the positive integers $\mathbb{Z}^+$ into infinitely many infinite subsets N_i, $\mathbb{Z}^+ = N_1 \cup N_2 \cup \cdots$, $N_i \cap N_j = \varnothing$. Let G_{N_i} be the group of eigenfunctions generated by $\{\omega_n(x) = \exp[2\pi i(x_n/p_n^2)] : n \in N_i\}$. Evidently the eigenvalues corresponding to functions in G_{N_i} form a dense subgroup of K, for every i. Let $f \in G_{N_i}$, $f \circ T = \alpha \cdot f$; then $F(x, y) = f(x)y$ satisfies $F(S^n(x, y)) = \alpha^n \exp[\pi i(\chi + \cdots + \chi \circ T^{n-1})] F(x, y)$. Hence with $\bar{m}$ the obvious $\mathbb{Z}_2$ extension of m

$$\int F \circ S^n \bar{F} \, d\bar{m} = \int \alpha^n z^n \, d\mu = \int z^n \, d\mu_\alpha,$$

where μ_α is the translation of μ by α. Since the set of α involved is dense in K, the maximal spectral type of U_S restricted to the subspace of $L^2(X \times \mathbb{Z}_2)$ spanned by $\{f(x)y : f \in G_{N_i}\}$ is Lebesgue measure, for each $i = 1, 2, \ldots$. Hence U_S has countable Lebesgue spectrum in the space spanned by $\{f(x)y : f$ is an eigenfunction of $T\}$. This latter space, however, is precisely the orthocomplement of $L^2(X)$.

We have to show that μ is absolutely continuous with respect to

Lebesgue measure where

$$r(n) = \int \exp\left[\pi i(\chi + \cdots + \chi \circ T^{n-1})\right] dm = \int_K z^n \, d\mu.$$

It is well known that it is sufficient to show $\sum_n |r(n)|^2 < \infty$.
Evidently,

$$r(n) = \int \exp\left[\pi i \sum_k S_k^n(x)\right] dm = \prod_k \int \exp\left[\pi i S_k^n(x)\right] dm$$

where $S_k^n(x) = \chi_{A_k}(x) + \cdots + \chi_{A_k}(T^{n-1}x)$, since m is the direct product measure and $S_k^n(x)$ depends only on the kth coordinate. We note that each integral (in the product) is at most 1 in absolute value so that we can write

$$|r(n)| \le \prod_{p_k^2 \ge 2n} |\int \exp\left[\pi i S_k^n(x)\right] dm|.$$

However, $S_k^n(x)$ is the characteristic function of $A_k \cup \cdots \cup T^{-(n-1)}A_k$
$= A_k^n$ for $n < p_k^2$. Hence

$$|r(n)| \le \prod_{p_k^2 \ge 2n} (-m(A_k^n) + 1 - m(A_k^n))$$

$$= \prod_{p_k^2 \ge 2n} [1 - (2n/p_k^2)]$$

$$\le \exp\left[-2n\left(\sum_{p_k^2 \ge 2n} 1/p_k^2\right)\right].$$

For some constant K we have $\sum_{p_k^2 \ge 2n} 1/p_k^2 \ge Kn^{-3/4}$. Hence $|r(n)| \le \exp(-2Kn^{1/4})$ and therefore $\sum_n |r(n)|^2 < \infty$, and the theorem is proved.

7 A weak-mixing transformation which is not strong-mixing

The following example is due to Kakutani and von Neumann (cf. Kakutani [2]). Let $(X, \mathscr{B}, m)$ be the usual Lebesgue space of the unit interval with its dyadic rationals removed. For $x \in X$ define

$$Tx = x + \tfrac{1}{2} \quad \text{if} \quad x < \tfrac{1}{2},$$
$$= x - \tfrac{1}{4} \quad \text{if} \quad \tfrac{1}{2} < x < \tfrac{3}{4},$$
$$= x - \tfrac{1}{4} - \tfrac{1}{8} \quad \text{if} \quad \tfrac{3}{4} < x < \tfrac{7}{8} \quad \text{etc.}$$

A picture of the transformation looks like this:

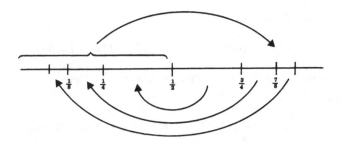

The transformation consists of piecewise translations of intervals. It is easy to show that T is ergodic with discrete spectrum consisting of eigenvalues $H(T) = \{\exp(2\pi i r) : r$ a dyadic rational$\}$. In fact, for each n, the intervals $(j/2^n, (j+1)/2^n)$, $j = 0, 1, \ldots, 2^n - 1$, are cyclically permuted by T and its iterates, though not in accordance with the order of the real line. Hence $\sum_{j=0}^{2^n - 1} \chi_{K_n} \circ T^j \omega_n^{-j}$ is an eigenfunction corresponding to the eigenvalue $\omega_n = \exp(2\pi i/2^n)$ and $K_n = (0, 1/2^n)$. It is not difficult to see that these eigenfunctions form an orthonormal basis for $L^2(X)$ and 'Fourier analysis' shows that T is ergodic with discrete spectrum. Now let $I_0 = (0, \frac{1}{2})$, $I_1 = (\frac{1}{2}, \frac{3}{4})$, $I_2 = (\frac{3}{4}, \frac{7}{8})$, ... and define $A = \bigcup_{n=0}^{\infty} I_{2n}$. Let A' be a copy of A (by the map $x \to x'$) disjoint from X, and define $X^A = X \cup A'$. X^A is provided with the obvious measure structure obtained from X and A' (as a copy of A). We define $T^A : X^A \to X^A$ by

$$\left. \begin{aligned} T^A(x') &= Tx \quad (x' \in A') \\ T^A(x) &= x' \quad (x \in A) \\ &= Tx \quad (x \in X - A). \end{aligned} \right\} \tag{5.5}$$

As Kakutani has shown, under more general circumstances, T^A is ergodic. In the case above this is not difficult to see by a consideration of orbits of T and of T^A. We show that T^A is weak-mixing but not strong-mixing.

Returning to the transformation T we shall need to know something of the form the function $\chi_A + \cdots + \chi_A \circ T^{2^n - 1}$ takes. Notice that each of the iterations $TJ_n, \ldots, T^{2^n - 1} J_n$, where $J_n = (1 - (1/2^n), 1)$ is either contained in A or is disjoint from A. J_n however contains a part of A and a part of A^c. $\chi_A + \cdots + \chi_A \circ T^{2^n - 1}$ is the number of visits x makes to A in the first 2^n iterations of T. This number would be independent of x, were it not for the visit to J_n. In other words, $S_n = \chi_A + \cdots + \chi_A \circ T^{2^n-1}$ assumes only two values, s_n and $s_n + 1$, where

$$m\{x : S_n(x) = s_n + 1\} = m(A) = \tfrac{2}{3}.$$

We note that $(T^A)^{1+\chi_A(x)}(x) = T(x)$ for all $x \in X$ and therefore $(T^A)^{2^n + S_n(x)}(x) = T^{2^n}(x)$ for all $x \in X$. Hence either $(T^A)^{2^n + s_n}(x) = (T^A)^{-1} T^{2^n}(x)$ or $T^{2^n}(x)$, and we conclude that

$$(T^A)^{2^n + s_n} I_0 \subset (T^A)^{-1} I_0 \cup I_0$$

or

$$(T^A)^{2^n + s_n + 1} I_0 \subset I_0 \cup T^A I_0 = I_0 \cup I_0'.$$

This certainly excludes strong-mixing for T^A since if B is a set of positive measure disjoint from $I_0 \cup I_0'$ then $m(T^A)^{2^n + s_n + 1}(I_0) \cap (B)$ does not tend to a positive number.

Finally we show that T^A is weak-mixing. Let f be an eigenfunction for T^A. We may assume $|f| = 1$, $f \circ T^A = \exp(2\pi i\alpha) \cdot f$ with $0 \le \alpha < 1$. In view of (5.5) we have $f \circ T = \exp[2\pi i\alpha(1 + \chi_A)] \cdot f$ and therefore $f(T_x^{2^n}) = \exp[2\pi i\alpha(2^n + S_n(x))] \cdot f(x)$. Hence

$$|\textstyle\int (f \circ T^{2^n}) \bar{f}\, dm| = |\textstyle\int \exp[2\pi i\alpha S_n(x)]\, dm|$$

and since $S_n(x) = s_n$ on a set of measure $1/3$ and $S_n(x) = s_n + 1$ on a set of measure $2/3$,

$$|\textstyle\int \exp[2\pi i\alpha S_n(x)]\, dm| = |\tfrac{1}{3} + \tfrac{2}{3}\exp(2\pi i\alpha)| < 1,$$

if $\alpha \ne 0$. But f can be approximated, in $L^2(X)$, arbitrarily closely by linear combinations of characteristic functions of dyadic intervals of length $1/2^n$ and these latter functions g have the property that $g \circ T^{2^n} = g$. This implies that $|\int (f \circ T^{2^n}) \bar{f}\, dm| \to 1$ so we must conclude that $\alpha = 0$ and f is constant. The proof that T^A is weak-mixing (but not strong-mixing) is complete.

Appendix

Spectral theory of unitary operators

1 Two basic theorems

We shall base our account on two theorems; the first is due to Herglotz (later generalised to locally compact abelian groups by Bochner (cf. Loomis [1])) and the second is due to Wiener (cf. Halmos [4]; Plessner and Rohlin [1] for general accounts of spectral theory).

A sequence $\{r_n\}$ of complex numbers is said to be *positive definite* if $\sum_{n, m = 0}^{N} r_{n - m} \, a_n \bar{a}_m \geq 0$ for all sequences $\{a_n\}$ and all non-negative integers N.

If U is a unitary operator on the Hilbert space H and $x \in H$, then $r_n = \langle U^n x, x \rangle \; (n = 0, \pm 1, \pm 2, \ldots)$ is positive definite since

$$\sum_{n, m = 0}^{N} \langle U^{n - m} x, x \rangle a_n \bar{a}_m = \left\langle \sum_{n = 0}^{N} a_n U^n x, \sum_{m = 0}^{N} a_m U^m x \right\rangle$$
$$\geq 0.$$

1. Theorem (Herglotz; cf. Feller [1]) *If $\{r_n\}$ is a positive definite sequence then there is a unique finite non-negative measure μ on $K = \{z : |z| = 1\}$ (or on $[0, 1)$) such that*

$$r_n = \int_K z^n \, d\mu = \int_0^1 \exp(2\pi i n x) \, d\mu.$$

Conversely, if μ is a non-negative measure on K then r_n defined as above is a positive definite sequence.

Proof Clearly $r_0 \geq 0$ and for all complex λ, $(1 + |\lambda|^2) r_0 + r_n \lambda + r_{-n} \bar{\lambda} \geq 0$. Hence $r_n \lambda + r_{-n} \bar{\lambda}$ is real for all complex λ and it is easy to see that $r_{-n} = \bar{r}_n$. Put $\lambda = \theta \bar{r}_n$ to obtain

$$(1 + |\theta|^2 |r_n|^2) r_0 + \theta |r_n|^2 + \bar{\theta} |r_n|^2 \geq 0$$

for all complex θ.

For real θ we have a quadratic in θ which is never negative. The discriminant of the quadratic shows that $|r_n| \leq r_0$ for all n. In particular, the sequence $\{r_n\}$ is bounded. We can dismiss the trivial case $r_0 = 0$, so that without loss of generality we take $r_0 = 1$. Let $0 < s < 1$, then positive definiteness yields

$$f_s(z) = \sum_{n,m=0}^{\infty} r_{n-m} s^{n+m} z^{m-n} \geq 0$$

for all $|z| = 1$. But this sum equals

$$\sum_{n=-\infty}^{\infty} r_n z^{-n} \sum_{m=0}^{\infty} s^{|n|+2m} = \sum_{n=-\infty}^{\infty} r_n z^{-n} s^{|n|} \frac{1}{1-s^2}.$$

Hence

$$\int_K f_s(z) z^{-n} \mathrm{d}z = \frac{r_{-n} s^{|n|}}{1-s^2}.$$

Define μ_s by

$$\frac{\mathrm{d}\mu_s}{\mathrm{d}z} = (1-s^2) f_s(z) \geq 0$$

so that

$$\int_K z^{-n} \mathrm{d}\mu_s = r_{-n} s^{|n|}, \quad \mu_s(K) = r_0 = 1.$$

Choose a sequence $s_m \to 1 \ (0 < s_m < 1)$; then $\int_K z^{-k} \mathrm{d}\mu_{s_m} \to r_{-k}$ for all $k = 0, \pm 1, \ldots$. Hence $\int_K p(z) \mathrm{d}\mu_{s_m}$ converges as $m \to \infty$ for all linear combinations $p(z)$ of functions z^k. Since such functions are dense in $C(K)$, we see that $\int_K f(z) \mathrm{d}\mu_{s_m}$ converges for all $f \in C(K)$ to, say, $J(f)$. But $\mathrm{d}\mu_s/\mathrm{d}z \geq 0$ implies $J(f) \geq 0$ when $f \geq 0$ and therefore $J(f) = \int f \mathrm{d}\mu$ for some probability μ on K. We conclude that

$$\int_K z^{-k} \mathrm{d}\mu = \lim_{m \to \infty} \int z^{-k} \mathrm{d}\mu_{sm} = r_{-k}$$

and the existence part of the theorem is complete.

It is clear that if $\int_K z^k \, \mathrm{d}\mu = r_k$ for some finite positive measure μ then $\{r_k\}$ is positive definite:

$$\sum_{n,m=0}^{N} r_{n-m} a_n \bar{a}_m = \sum_{n,m=0}^{N} a_n \bar{a}_m \int_K z^{n-m} \mathrm{d}\mu$$

$$= \int_K \left| \sum_{n=0}^{N} a_n z^n \right|^2 \mathrm{d}\mu \geq 0.$$

Finally the measure μ such that $\int_K z^k \, \mathrm{d}\mu = r_k$ is unique since $\int_K z^k \, \mathrm{d}\mu$

$= \int_K z^k \, dv$ for all $k = 0, \pm 1, \ldots$ implies $\mu \equiv v$. We simply note that $\int_K f(z) \, d\mu = \int_K f(z) \, dv$ for all finite linear combinations of $\{z^k\}$ and therefore for all $f \in C(K)$.

2. Theorem (Wiener) *Let m be a finite Borel measure defined on the circle K.*

If H is a closed subspace of $L^2(K, m)$ which is invariant with respect to the unitary operator $V : f(z) \rightarrow z f(z)$ (i.e. $VH = H$) then

$$H = \chi_B L^2(K, m) = \{ f \in L^2(K, m) : f = 0 \quad \text{on } B^c \}$$

for some Borel subset B.

Proof Let $1 = k + h$ be the orthogonal decomposition of 1 with respect to $H^\perp, H$ (i.e. $k \in H^\perp, h \in H$). Then $k \perp V^n h$ for all n; $\int k(z) \cdot \overline{h(z)} z^n \, dm = 0$, $n = 0, \pm 1, \ldots$. Therefore $k \, \bar{h} = 0$ (a.e.) and $1 = |k|^2 + |h|^2$ (a.e.). Since k, h have disjoint 'supports' ($k = \chi_A k, h = \chi_{A^c} h$), $|k| = 1$ on A and $|h| = 1$ on A^c. But $1 = k + h$ implies $k = 1$ on A, $h = 1$ on A^c. In other words $1 = \chi_A + \chi_{A^c}$ is the decomposition of 1 with respect to $H^\perp, H$. Hence $z^n \chi_{A^c}(z) \in H$ for $n = 0, \pm 1, \ldots$ and we conclude that $\chi_{A^c} L^2(K, m) \subset H$, $\chi_A L^2(K, m) \subset H^\perp$ i.e. $\chi_{A^c} L^2(K, m) = H$.

2 Spectral multiplicity theorems

If $U_i (i = 1, 2)$ are isometries of the Hilbert spaces H_i then U_1 is said to be *unitarily* (or *spectrally*) *equivalent* to U_2 if there exists an isometry W of H_1 onto H_2 such that $W U_1 = U_2 W$. In this case we write $U_1 \simeq U_2$.

We wish to characterise isometries up to unitary equivalence. The first thing to note is that if U is an isometry acting on the Hilbert space H then $U|H_\infty$ is a unitary operator where $H_\infty = \bigcap_{n=0}^\infty U^n H$, since $U H_\infty = H_\infty$ (i.e. U is invertible). $U|H_\infty^\perp$ is very easy to characterise. In fact $H_\infty^\perp = V \oplus UV \oplus \cdots$ where $V = H \ominus UH$, and therefore $U|H_\infty^\perp$ is completely described (up to unitary equivalence) by the dimension of V. *For this reason we shall be concerned exclusively with unitary operators U on a Hilbert space H.*

Let U be a unitary operator on the Hilbert space H. For $x \in H$, $Z(x)$ denotes the *cycle* (or *cyclic subspace*) *generated by* x which is the closure of the linear span of $\{ U^n x : n = 0, \pm 1, \ldots \}$.

U|Z(x) is unitarily (spectrally) equivalent to

$$V_x : L^2(K, \tilde{x}) \to L^2(K, \tilde{x}) \text{ defined by } (V_x f)(z) = z f(z) \quad (A.1)$$

where $\tilde{x}$ is the *spectral measure* of x (with respect to U) i.e. $\langle U^n x, x \rangle$ $= \int_K z^n \, d\tilde{x}$ for all n.

In fact, if we define $W : U^n x \to z^n \in L^2(K, \tilde{x})$ then W is an isometry on $\{U^n x : n = 0, \pm 1, \ldots\}$ since $\langle U^m x, U^n x \rangle = \int z^m z^{-n} \, d\tilde{x}$. Hence W extends to an isometry of $Z(x)$ onto $L^2(K, \tilde{x})$. Clearly $WU = V_x W$, and this is what we mean by the unitary equivalence of $U|Z(x)$ and V_x.

U|Z(x) is unitarily equivalent to U|Z(y)

$$(U|Z(x) \simeq U|Z(y)) \text{ if and only if } \tilde{x} \simeq \tilde{y}. \quad (A.2)$$

We have to show that $V_x \simeq V_y$ if and only if $\tilde{x} \simeq \tilde{y}$. Suppose $WV_x = V_y W$ for some isometry W and write $f(z) = W(1)$, then $WV_x^n 1 = V_y^n f$, i.e. $W(z^n) = f(z)z^n$. Hence W is the multiplication operator $g \to f \cdot g$ and χ_B in $L^2(K, \tilde{x})$ has the same norm as $f \cdot \chi_B$ in $L^2(K, \tilde{y})$, i.e. $\tilde{x}(B) = \int_B |f|^2 \, d\tilde{y}$. Therefore $\tilde{x} \leq \tilde{y}$. A similar argument shows that $\tilde{y} \leq \tilde{x}$ and hence $\tilde{x} \simeq \tilde{y}$.

On the other hand, if $\tilde{x} \simeq \tilde{y}$ define $W : L^2(K, \tilde{x}) \to L^2(K, \tilde{y})$ by $Wg = g \cdot (d\tilde{x}/d\tilde{y})^{1/2}$. W is an isometry and $WV_x = V_y W$.

If $x \in H$ and $\mu \leq \tilde{x}$ is a finite non-negative Borel measure on K then there exists $y \in Z(x)$ with $\tilde{y} = \mu$. (A.3)

It suffices to note the existence of $f \in L^2(K, \tilde{x})$ with $\tilde{f} = \mu$. In fact $f = (d\mu/d\tilde{x})^{1/2}$ satisfies

$$\langle V_x^n f, f \rangle = \int z^n \frac{d\mu}{d\tilde{x}} \cdot d\tilde{x} = \int z^n \, d\mu = \int z^n \, d\tilde{f}$$

for all n, i.e. $\mu = \tilde{f}$.

If $x, y \in Z(z)$ and $Z(x) \perp Z(y)$ then $\tilde{x} \perp \tilde{y}$.

If in addition $z = x + y$ then $Z(z) = Z(x) \oplus Z(y)$. (A.4)

Transferring to $L^2(K, \tilde{z})$ we show that $Z(f) \perp Z(g)$, $f, g \in L^2(K, \tilde{z})$ implies $\tilde{f} \perp \tilde{g}$. In fact $Z(f) = \chi_A \cdot L^2(K, \tilde{z})$ and $Z(g) = \chi_B L^2(K, \tilde{z})$ (by Wiener's theorem) and orthogonality ensures that $\tilde{z}(A \cap B) = 0$. Since $d\tilde{f} = |f|^2 d\tilde{z}$, $d\tilde{g} = |g|^2 d\tilde{z}$ we have $\tilde{f} \perp \tilde{g}$. If we assume now that $z = x + y$ then $1 = f + g$ and $Z(1) = L^2(K, \tilde{z}) = Z(f) + Z(g)$.

If $y \in Z(x)$ then $\tilde{y} \leq \tilde{x}$ with equivalence holding

when and only when $Z(y) = Z(x)$.   (A.5)

Map $Z(x)$ to $L^2(K, \tilde{x})$ (by sending $U^n x$ to z^n) and let f denote the image of y. Then we have to show that $\tilde{f} \leq \tilde{x}$ with equivalence holding when and only when $Z(f) = Z(1)$ (with respect to V_x). But $\langle V_x^n f, f \rangle$ $= \int z^n d\tilde{f} = \int z^n |f|^2 d\tilde{x}$. Hence $d\tilde{f} = |f|^2 d\tilde{x} \leq d\tilde{x}$. If $Z(y) = Z(x)$ then $U|Z(y) \simeq U|Z(x)$ and we have seen that $\tilde{x} \simeq \tilde{y}$. If $Z(y)$ is a proper subspace of $Z(x)$ then $Z(f)$ is a proper subspace of $L^2(K, \tilde{x})$ invariant under V_x. By Wiener's theorem $Z(f) = \chi_B L^2(K, \tilde{x})$ where $\tilde{x}(B) < \tilde{x}(K)$ and hence $\tilde{x}(B^c) > 0$, $\tilde{f}(B^c) = 0$ i.e. $\tilde{y}$, $\tilde{x}$ are not equivalent.

If $\tilde{x} \perp \tilde{y}$ (mutually singular) then $Z(x) \perp Z(y)$.   (A.6)

The converse is *not* always true. It is this fact which gives rise to multiplicity.

Write $y = y_0 + y_1$, with $y_1 \in Z(x)$, $y_0 \perp Z(x)$ so that $Z(y_0) \perp Z(x)$. $\langle U^n y, y \rangle = \langle U^n y_0, y_0 \rangle + \langle U^n y_1, y_1 \rangle$, that is $\int z^n d\tilde{y} = \int z^n d\tilde{y}_0$ $+ \int z^n d\tilde{y}_1$. Hence $\tilde{y} = \tilde{y}_0 + \tilde{y}_1 \perp \tilde{x}$. But $y_1 \in Z(x)$ implies $\tilde{y}_1 \leq \tilde{x}$. Therefore $\tilde{y}_1 = 0$ and we conclude that $y_1 = 0$, $y = y_0$ and $Z(x) \perp Z(y)$.

If $\tilde{x} \perp \tilde{y}$ then $\widetilde{x+y} = \tilde{x} + \tilde{y}$ and

$$Z(x + y) = Z(x) \oplus Z(y).$$ (A.7)

Since $Z(x) \perp Z(y)$,

$$\langle U^n(x + y), (x + y) \rangle = \langle U^n x, x \rangle + \langle U^n y, y \rangle,$$

i.e.

$$\int z^n d\widetilde{x+y} = \int z^n d\tilde{x} + \int z^n d\tilde{y} \quad \text{so that} \quad \widetilde{x+y} = \tilde{x} + \tilde{y}.$$

Now $d\tilde{x}/d\widetilde{x+y} \in L^2(K, \widetilde{x+y})$ so that for $\varepsilon > 0$ there exists a polynomial p in z, z^{-1} with

$$\int \left| \frac{d\tilde{x}}{d\widetilde{x+y}} - p(z) \right|^2 d\widetilde{x+y} < \varepsilon.$$

Hence $\|x - p(U)(x + y)\|^2$

$$= \langle x, x \rangle - 2\mathcal{R}\langle x, p(U)(x + y) \rangle + \|p(U)(x + y)\|^2$$

$$= \int 1 d\tilde{x} - 2\mathcal{R}\langle x, p(U)x \rangle + \int |p(z)|^2 d\widetilde{x+y}$$

$$= \int 1 d\tilde{x} - 2\mathcal{R} \int p(z) d\tilde{x} + \int |p(z)|^2 d\widetilde{x+y}$$

$$= \int 1 d\tilde{x} - \int \frac{d\tilde{x}}{d\widetilde{x+y}} d\tilde{x} + \int \left| \frac{d\tilde{x}}{d\widetilde{x+y}} - p(z) \right|^2 d\widetilde{x+y}$$

$$= \int \left| \frac{d\tilde{x}}{d\widetilde{x+y}} - p(z) \right|^2 d\widetilde{x+y} < \varepsilon.$$

Since $\varepsilon > 0$ is arbitrary $x \in Z(x + y)$. In the same way $y \in Z(x + y)$. Therefore $Z(x + y) \supset Z(x) \oplus Z(y)$. If $v \in Z(x + y)$ and $v \perp Z(x) \oplus Z(y)$ then for $\varepsilon > 0$ there exists a polynomial p in z, z^{-1} with $\|v - p(U)(x + y)\|^2 < \varepsilon$. Hence $\|v\|^2 + \|p(U)(x + y)\|^2 < \varepsilon$. Since $\varepsilon > 0$ is arbitrary, $v = 0$. Hence $Z(x + y) = Z(x) \oplus Z(y)$.

A cyclic subspace $Z(x)$ is said to be *maximal* if it is contained in no larger cyclic subspace. Evidently $Z(x)$ is maximal if and only if $\tilde{x} \geq \tilde{y}$ for all $y \in H$. $\tilde{x}$ is then called (along with other $\tilde{y} \sim \tilde{x}$) a *maximal spectral type*.

If U is a unitary operator of a separable Hilbert space H,

then there exists a maximal cyclic subspace. In fact if $x \in H$,

there is a maximal cyclic subspace containing x. (A.8)

This is easily proved by using Zorn's lemma applied to all cyclic subspaces containing x.

If U_i are unitary operators on H_i ($i = 1, 2$) such

that $U_1 \simeq U_2$ and $U_1 | Z(x) \simeq U_2 | Z(y)$ then

$$U_1 | Z(x)^{\perp} \simeq U_2 | Z(y)^{\perp}.$$ (A.9)

The problem can be transferred to one space. It becomes: if $U | Z(x) \simeq U | Z(y)$ then $U | Z(x)^{\perp} \simeq U | Z(y)^{\perp}$. It will suffice to show that

$$U | \overline{Z(x) + Z(y)} \ominus Z(x) \simeq U | \overline{Z(x) + Z(y)} \ominus Z(y)$$

since $U | \overline{Z(x) + Z(y)} \simeq U | \overline{Z(x) + Z(y)}$ (using the identity). In other words we may assume $\overline{Z(x) + Z(y)} = H$. Let $y = y_0 + y_1$, $y_0 \perp Z(x)$, $y_1 \in Z(x)$; then $H = Z(x) \oplus Z(y_0)$. Write $Z(x) = Z(x_0) \oplus Z(y_1)$ (some x_0); then

$$H = \overline{Z(x) + Z(y)} = Z(x_0) \oplus Z(y_0) \oplus Z(y_1).$$

$$\tilde{x}_0 + \tilde{y}_1 = \tilde{x} \sim \tilde{y} = \tilde{y}_0 + \tilde{y}_1, \quad \tilde{x}_0 \perp \tilde{y}_1, \quad \tilde{y}_0 \perp \tilde{y}_1$$

and therefore $\tilde{x}_0 \sim \tilde{y}_0$. We conclude that there is an isometry from $Z(x_0)$ onto $Z(y_0)$ conjugating $U | Z(x_0)$ to $U | Z(y_0)$ and the proof is easily completed.

If U is a unitary operator on a separable Hilbert space

H then H can be decomposed into an orthogonal sum of

cyclic subspaces $H = \sum_n \oplus Z(x_n)$ with $\tilde{x}_1 \geq \tilde{x}_2 \geq \cdots$. (A.10)

Such a decomposition is called a *canonical* decomposition into 'decreasing' cycles. In fact let $\{y_n\}$ be dense in H. Choose a maximal cyclic space containing y_1 – say $Z(x_1)$. Now let $\{y_n^1 : n = 2, 3, \ldots\}$ be the projections of $\{y_n : n = 2, 3, \ldots\}$ onto $Z(x_1)^{\perp}$. Choose a maximal cyclic subspace (with respect to $U|Z(x_1)^{\perp}$) of $Z(x_1)^{\perp}$ containing y_2^1 – say $Z(x_2)$. Note that $y_1 \in Z(x_1)$, $y_2 \in Z(x_2) \oplus Z(x_1)$ (part is in $Z(x_1)$ and part in $Z(x_2)$). Repeat this process indefinitely. This process ensures that $y_n \in Z(x_1) \oplus \cdots \oplus Z(x_n)$. Since the $\{y_n\}$ are dense, $H = \sum_n \oplus Z(x_n)$. By maximality $\tilde{x}_1 \geq \tilde{x}_2 \cdots$.

The decomposition above is summarised by $\tilde{x}_1$ and the sets A_2, $A_3, \ldots$ where A_2 is the support of $d\tilde{x}_2/d\tilde{x}_1$ (with respect to $\tilde{x}_1$) etc., $K = A_1 \supset A_2 \supset \cdots$. Hence the decomposition is summarised by $\tilde{x}_1 = \mu_U$ and $M_U = \sum_{n=1}^{\infty} \chi_{A_n}$. M_U is called the *multiplicity function* and is defined a.e. with respect to μ_U.

3. Theorem *If* $U_1|H_1 \simeq U_2|H_2$ *then* $\mu_{U_1} \sim \mu_{U_2}$ *and* $M_{U_1} = M_{U_2}$.

We need only show that canonical decompositions $H_1 = \sum_{n=1}^{\infty} \oplus Z(x_n)$, $H_2 = \sum_{n=1}^{\infty} \oplus Z(y_n)$ satisfy $\tilde{x}_1 \sim \tilde{y}_1, \tilde{x}_2 \sim \tilde{y}_2, \ldots$. We know $\tilde{x}_1 \sim \tilde{y}_1$. Now consider $U_1|Z(x_1)^{\perp}$ and $U_2|Z(y_1)^{\perp}$ which we know to be spectrally equivalent. Clearly $Z(x_1)^{\perp} = \sum_{n=2}^{\infty} \oplus Z(x_n)$ and $Z(y_1)^{\perp} = \sum_{n=2}^{\infty} \oplus Z(y_n)$. Again we know $\tilde{x}_2 \sim \tilde{y}_2$. The proof is easily completed by induction.

4. Theorem *If* $U_1|H_1$ *and* $U_2|H_2$ *have maximal spectral types* $\mu_{U_1} \sim \mu_{U_2}$ *and multiplicities* $M_{U_1} = M_{U_2}$, *then* $U_1 \simeq U_2$.

This follows from the obvious fact: if $U|H$ has a canonical decomposition $H = \sum_{n=1}^{\infty} \oplus Z(x_n)$ then $U \simeq \sum_{n=1}^{\infty} \oplus V_{x_n} \simeq \sum_{n=1}^{\infty} \oplus V_{\mu_n}$ where $d\mu_n/d\mu_1 = \chi_{A_n}$, $\mu_n \sim \tilde{x}_n$.

3 Decompositions

Let U be a unitary operator on the Hilbert space H. An eigenvector is an element $x \in H$ such that $Ux = \lambda x$ for some $\lambda \in \mathbb{C}$; λ is the eigenvalue corresponding to $x(|\lambda| = 1)$. The discrete spectrum subspace of H, denoted by V_d, is the closure of the linear span of all eigenvectors. Clearly $UV_d = V_d$. Each $x \in V_d$ can be written as an orthogonal sum $x = \sum a(i)x_i$ with x_i an eigenvector with, say, eigenvalue λ_i. Hence

$U^n x = \sum a(i)\lambda_i^n x_i$ and

$$\langle U^n x, x \rangle = \sum |a(i)|^2 \, \lambda_i^n \|x_i\|^2 = \int_K z^n \, d\tilde{x}$$

so that $\tilde{x}$ is the *atomic* measure which assigns measure $|a(i)|^2 \, \|x_i\|^2$ to $\lambda_i \in K$.

On the other hand, if $x \in H$ and if $\tilde{x}$ is purely atomic, then for each $\lambda_i \in K$ with $\tilde{x}(\lambda_i) > 0$ we have a measure μ_i such that $\mu_i(\lambda_i) = \tilde{x}(\lambda_i)$ and $\mu_i(K - \{\lambda_i\}) = 0$. Hence $\mu_i \leq \tilde{x}$ and $\mu_i = \tilde{x}_i$ for some $\tilde{x}_i \in Z(x)$. Since $\tilde{x}_i$ is concentrated on the single point λ_i, we have

$$\langle U^n x_i, x_i \rangle = \int z^n \, dx_i = \lambda_i^n x(\lambda_i) = \lambda_i^n \|x_i\|^2.$$

The converse of Schwarz's inequality shows that $U x_i = \lambda_i x_i$. Moreover $\tilde{x} = \sum \tilde{x}_i$ and therefore $x \in V_d$. We have proved

For each $x \in H$, x is purely atomic if and only if $x \in V_d$.

As an immediate consequence we have

For each $x \in H$, $\tilde{x}$ is a non-atomic (i.e. continuous) measure if and only if $x \in V_c = V_d^\perp$.

***Exercise* 1**   Let $V_c = V_l \oplus V_l^\perp$ where $U V_l = V_l$ and $x \in V_l$ if and only if $\tilde{x}$ is absolutely continuous with respect to Lebesgue measure. Show that $x \in V_l^\perp$ if and only if $\tilde{x}$ is a continuous singular (with respect to Lebesgue) measure.

***Exercise* 2**   Show that $\tilde{x}$ is equivalent to Lebesgue measure if and only if there exists $y \in Z(x)$ with $Z(x) = Z(y)$ and $\langle U^n y, y \rangle = 0$ for all $n \neq 0$.

***Exercise* 3**   We say that x is a weak-mixing vector if $(1/N)\sum_{n=0}^{N-1} |\langle U^n x, x \rangle| \to 0$. Show that x is weak-mixing if and only if $\tilde{x}$ is non-atomic or, equivalently, $(\tilde{x} \times \tilde{x})(D) = 0$ where $D = \{(\lambda, \lambda) : \lambda \in K\} \subset K \times K$.

References

Abramov, L. M. [1]. Metric automorphisms with quasi-discrete spectrum. *Izv. Akad. Nauk. SSSR. Ser. Mat.* **26** (1962), 513–30 (Russian) – *Amer. Math. Soc. Transl. Ser.* 2 **39** (1964), 37–56.

Aczél, J. and Daróczy, Z. [1]. *On measures of information and their characterisations.* Academic Press, New York. 1975.

Adler, R. and Weiss, B. [1]. *Similarity of automorphisms of the torus.* Memoirs A.M.S. no. 98. 1970. Providence, R.I.

Ambrose, W. [1]. Representations of ergodic flows. *Ann. Math.* **42** (1941), 723–39.

Ambrose, W., Halmos, P. R. and Kakutani S. [1]. The decomposition of measures II. *Duke Math. J.* **9** (1942), 43–7.

Ambrose, W. and Kakutani, S. [1]. Structure and continuity of measurable flows. *Duke Math. J.* **9** (1942), 15–42.

Andronov, A. A. and Pontrjagin, L. S. [1]. Systèmes grossiers. *Comptes Rendus de l'Academie des Sciences de l'URSS* **14** (1937), 247–50.

Anosov, D. V. [1]. Geodesic flows on closed Riemannian manifolds with negative curvature. *Trudy. Mat. Inst. Steklova* **90** (1967), 1–209 (Russian) – *Proceedings of the Steklov Institute of Mathematics* (*Amer. Math. Soc. Transl.*) (1969), 1–235.

Anzai, H. [1]. Ergodic skew transformations on the torus. *Osaka Math. J.* **3** (1951), 83–99.

Arnold, V. I. [1]. Small denominators. I. Mapping the circle onto itself. *Izv. Akad. Nauk. SSSR. Ser. Mat.* **25** (1961), 21–86; correction **28** (1964), 479–480 (Russian) – *Amer. Math. Soc. Transl. Ser.* 2 **46** (1965), 213–84.

Arnold, V. I. [2]. On the classical theory of perturbations and the problem of stability of planetary systems. *Dokl. Akad. Nauk. SSSR* **145** (1962), 487–90 (Russian) – *Sov. Math. Dokl.* **3** (1962), 1008–12.

Arnold, V. I. [3]. Proof of a theorem of A. N. Kolmogorov on the preservation of conditionally periodic motions under a small perturbation of the Hamiltonian. *Usp. Mat. Nauk.* **18** (1963), 13–40 (Russian) – *Russian Math. Surveys* **18** (1963), 9–36.

Arnold, V. 1. [4]. Small denominators and problems of stability of motion in classical and celestial mechanics. *Usp. Mat. Nauk.* **18** (1963), 91–192 (Russian) – *Russian Math. Surveys* **18** (1963), 85–193.

Arnold, V. I. and Avez, A. [1]. *Ergodic problems of classical mechanics.* Mathematical physics monograph series. Benjamin, New York. 1968.

Arnold, V. I. and Sinai, Ja. G. [1]. On small perturbations of automorphisms of the torus. *Dokl. Akad. Nauk. SSSR.* **144** (1962), 695–698 (Russian) – *Sov. Math. Dokl.* **3** (1962), 783–6. Correction *Dokl. Akad. Nauk. SSSR.* **150** (1963), 958 – *Sov. Math. Dokl.* **4** (1963), (3), vi.

Auslander, L., Green, L. and Hahn, F. [1]. *Flows on homogeneous spaces. Annals of Math. Studies* **53**. Princeton. 1963.

Billingsley, P. [1]. *Ergodic theory and information.* Wiley, New York. 1965.

Birkhoff, G. D. [1]. Proof of the ergodic theorem. *Proc. Nat. Acad. Sci. U.S.A.* **17** (1931), 656–60.

Birkhoff, G. D. [2]. *Dynamical Systems.* A.M.S. Colloquium Publications no. 9, Providence. 1927. (Revised 1966.)

Birkhoff, G. D. and Koopman, B. O. [1]. Recent contributions to the ergodic theory. *Proc. Nat. Acad. Sci. U.S.A.* **18** (1932), 279–82.

Bowen, R. [1]. Markov partitions for Axiom A. diffeomorphisms. *Amer. J. Math.* **92** (1970), 725–47.

Bowen, R. [2]. *Equilibrium states and the ergodic theory of Anosov diffeomorphisms.* Lecture notes in mathematics, no. 470, Springer, Berlin. 1975.

Breiman, L. [1]. The individual ergodic theorem of information theory. *Ann. Math. Stat.* **28** (1957) 809–11. Correction, *Ann. Math. Stat.* **31** (1960), 809–10.

Chacon, R. V. [1]. Change of velocity in flows. *J. Math. Mech.* **16** (1966), 417–31.

Chacon, R. V. and Ornstein, D. S. [1]. A general ergodic theorem. *Ill. Math. J.* **4** (1960), 153–60.

Chung, K. L. [1]. A note on the ergodic theorem of information theory. *Ann. Math. Stat.* **32** (1961), 612–14.

Chung, K. L. [2]. *Markov chains with stationary transition probabilities.* Springer, Berlin. 1960.

Dani, S. G. [1]. Bernoullian translations and minimal horospheres on homogeneous spaces. *J. Indian Math. Soc.* **40** (1978), 245–84.

Doob, J. L. [1]. *Stochastic processes.* Wiley, New York. 1953.

Dowker, Y. N. [1]. Invariant measure and the ergodic theorems. *Duke Math. J.* **14** (1947), 1051–61.

Dowker, Y. N. [2]. Finite and σ-finite invariant measures. *Ann. Math.* **54** (1951), 595–608.

Dunford, N. and Schwartz, J. T. [1]. *Linear Operators I.* Interscience, John Wiley. New York. Part 1, 1958.

Ellis, R. [1]. *Lectures on topological dynamics.* Benjamin, New York. 1969.

Feldman, J. [1]. New K-automorphisms and a problem of Kakutani. *Israel J. Math.* **24** (1976), 16–38.

Feller, W. [1]. *An introduction to probability theory and its applications* vols. I and II. Wiley, New York. 1957 and 1966.

Friedman, N. [1]. *Introduction to ergodic theory.* Van Nostrand Reinhold Studies no. 29, New York. 1970.

Friedman, N. and Ornstein, D. [1]. *Entropy and the isomorphism problem* (Preprint.)

Friedman, N. and Ornstein, D. [2]. On the isomorphism of weak Bernoulli transformations. *Adv. in Math.* **5** (1970), 365–94.

Furstenberg, H. [1]. Strict ergodicity and transformations of the torus. *Amer. J. Math.* **83** (1961), 573–601.

Gallovotti, G. [1]. Ising model and Bernoulli schemes in one dimension. *Comm. Math. Phys.* **32** (1973), 183–90.

Garsia, A. [1]. A simple proof of E. Hopf's maximal ergodic theorem. *J. Math. Mech.* **14** (1965), 381–2.

Garsia, A. [2]. *Topics in almost everywhere convergence* (Lectures in advanced mathematics), Markham, Chicago, 1970.

Gibbs, J. W. [1]. *Elementary principles in statistical mechanics.* Dover, New York. 1960.

Gottschalk, W. H. and Hedlund, G. A. [1]. *Topological dynamics.* A.M.S. Colloquium publications, no. 36, Providence, 1955.

Hahn, F.˙J. [1]. On an affine transformation of compact abelian groups. *Amer. J. Math.* **85** (1963), 428–6. Errata **86** (1964), 463–4.

Hajian, A. and Kakutani, S. [1]. Weakly wandering sets and invariant measures. *Trans. Amer. Math. Soc.* **110** (1964), 136–51.

Halmos, P. R. [1]. *Lectures on ergodic theory.* Chelsea, New York. 1958.

Halmos, P. R. [2]. The decomposition of measures. *Duke Math. J.* **8** (1941), 386–92.

Halmos, P. R. [3]. Measurable transformations. *Bull. Amer. Math. Soc.* **55** (1949), 1015–34.

Halmos, P. R. [4]. *Introduction to Hilbert spaces.* Chelsea, New York. 1957.

Halmos, P. R. and von Neumann, J. [1]. Operator methods in classical mechanics. II. *Ann. Math.* **43** (1942), 332–50.

Helson, H. and Kahane, J. P. [1]. Compact groups with ordered duals III. *J. London Math. Soc.* **4** (1972), 573–5.

Helson, H. and Parry, W. [1]. Cocycles and spectra. *Arkiv för Matematik* **16** (1978), 195–206.

Herman, M. R. [1]. *Thèses présentées à l'Université Paris Sud Centre d'Orsay,* 1976.

Hopf, E. [1]. *Ergodentheorie.* Chelsea, New York. 1948.

Hopf, E. [2]. On the time average theorem in dynamics. *Proc. Nat. Acad. Sci. U.S.A.* **18** (1932), 93.

Hopf, E. [3]. On the ergodic theorem for positive linear operators. *J. Math. Mech.* **205** (1961), 101–6.

Hopf, E. [4]. Theory of measures and invariant integrals. *Trans. Amer. Math. Soc.* **34** (1932), 373–3.

Humphreys, P. D. [1]. Change of velocity in dynamical systems. *J. London Math. Soc.* (2), **7** (1974), 747–57.

Hurewicz, W. [1]. Ergodic theorem without invariant measure. *Ann. Math.* **45** (1944), 192–206.

Ionescu Tulcea, A. [1]. Contributions to information theory for abstract alphabets. *Arkiv. Mat.* **4** (1960) 235–47.

Jacobs, K. [1]. *Neuere Methoden und Ergebnisse der Ergoden-theorie.* Springer, Berlin. 1960.

Kakutani, S. [1]. Ergodic theory. *Proc. Int. Cong. Math. Cambridge* (*Mass.*) 1950 **2** (1952), 128–42.

Kakutani, S. [2]. Examples of ergodic measure-preserving transformations which are weakly mixing but not strongly mixing. *Recent advances in topological dynamics* pp. 143–9. Lecture notes in mathematics no. 318. Springer, Berlin. 1973.

Kakutani, S. and Yosida, K. [1]. Birkhoff's ergodic theorem and the maximal ergodic theorem. *Proc. Imp. Acad. Tokyo.* 15 (1939), 165–8.

Katznelson, Y. [1]. Ergodic automorphisms of T^n are Bernoulli shifts. *Israel J. Math.* 10 (1971), 186–95.

Kendall, D. G. [1]. Functional equations in information theory. *Z. Wahrscheinlichkeitstheorie verw. Geb.* 2 (1963), 225–9.

Khintchine, A. [1]. Zur Birkhoff's Losung des Ergoden-problems. *Math. Ann.* 107 (1932), 485–8.

Khintchine, A. [2]. Eine Verschärfung des Poincaréschen Wiederkehrsatzes. *Comp. Math.* 1 (1934), 177–9.

Kolmogorov, A. N. [1]. La théorie générale des systèmes dynamiques et la mécanique classique. *Int. Cong. Math. Amsterdam* 1 (1954), 315–3.

Kolmogorov, A. N. [2]. The conservation of conditionally periodic motions under small perturbations of the Hamiltonian. *Dokl. Akad. Nauk. SSSR.* 98 (1954), 527–30 (Russian).

Kolmogorov, A. N. [3]. A new metric invariant of transient dynamical systems and automorphisms of Lebesgue spaces. *Dokl. Akad. Sci. SSSR.* 119 (1958), 861–64 (Russian).

Kolmogorov, A. N. [4]. On dynamical systems with an integral invariant on the torus. *Dokl. Akad. Nauk. SSSR.* 93 (1953), 763–6.

Koopman, B. O. [1]. Hamiltonian systems and linear transformations in Hilbert space. *Proc. Nat. Acad. Sci. U.S.A.* 17 (1931), 315–18.

Krylov, N. and Bogolioubov, N. [1]. La théorie générale de la mesure dans son application à l'étude des systèmes dynamiques de la mécanique non-linéaire. *Ann. Math.* 38 (1937), 65–113.

Lee, P. M. [1]. On the axioms of information theory. *Ann. Math. Statist.* 35 (1964), 415–18.

Loomis, L. H. [1]. *An introduction to abstract harmonic analysis.* Van Nostrand, Princeton. 1953.

Mackey, G. W. [1]. Ergodic theory and its significance for statistical mechanics and probability theory. *Adv. in Math.* **12** (1974), 178–268.

McMillan, B. [1]. The basic theorem of information theory. *Ann. Math. Stat.* **24** (1953), 196–219.

Miles, G. and Thomas, R. K. [1]. (i) Generalised torus automorphisms are Bernoulli. (ii) On the polynomial uniformity of translations of the *n*-torus. (iii) The breakdown of automorphisms of compact topological groups. (3 preprints.)

Moser, J. [1]. On invariant curves of area preserving mappings of an annulus. *Nachr. Akad. Wiss. Göttingen Math. Phys. Kl. Ila.* **1** (1962), 1–20.

Moser, J., Phillips, E. and Varadhan, S. [1]. *Ergodic theory (a seminar)* Courant Institute, New York University. 1975.

Nemytskii, V. and Stepanov, V. V. [1]. *Qualitative theory of differential equations.* Princeton. 1960.

Newton, D. [1]. Coalescence and spectrum of automorphisms of a Lebesgue space. *Z. Wahrscheinlichkeitstheorie verw. Geb.* **19** (1971), 117–22.

Ornstein, D. S. [1]. On invariant measures. *Bull. Amer. Math. Soc.* **66** (1960), 297–300.

Ornstein, D. S. [2]. *Ergodic theory, randomness and dynamical systems.* Yale University Press. 1974.

Ornstein, D. S. [3]. Bernoulli shifts with the same entropy are isomorphic. *Adv. in Math.* **4** (1970), 337–52.

Ornstein, D. S. [4]. A K-automorphism with no square root and Pinsker's conjecture. *Adv. in Math.* **10** (1973), 89–102.

Ornstein, D. S. [5]. *Imbedding Bernoulli shifts in flows.* Lecture notes in mathematics no. 160, pp. 178–218. Springer, Berlin, 1970.

Ornstein, D. S. [6]. Some new results in the Kolmogorov–Sinai theory of entropy and ergodic theory. *Bull. Amer. Math. Soc.* **77** (1971), 878–90.

Ornstein, D. S. [7]. The isomorphism theorem for Bernoulli flows. *Adv. in Math.* **10** (1973), 124–142.

Ornstein, D. S. and Shields, P. [1]. An uncountable family of K-automorphisms. *Adv. in Math.* **10** (1973), 63–88.

Ornstein, D. S. and Weiss, B. [1]. Geodesic flows are Bernoullian. *Israel J. Math.* **14** (1973), 184–98.

Oxtoby, J. C. [1]. Ergodic sets. *Bull. Amer. Math. Soc.* **58** (1952), 116–136.

Parry, W. [1]. Ergodic properties of affine transformations and flows on nilmanifolds. *Amer. J. Math.* **91** (1969), 757–71.

Parry, W. [2]. Metric classification of ergodic nilflows and unipotent affines. *Amer. J. Math.* **93** (1971), 819–28.

Parry, W. [3]. *Entropy and generators in ergodic theory.* Benjamin, New York. 1969.

Parry, W. [4]. Cocycles and velocity changes (2) *J.L.M.S.* **5** (1972), 511–16.

Peixoto, M. M. [1]. On structural stability. *Ann. Math.* **69** (1959), 199–222.

Pinsker, M. S. [1]. *Information and information stability of random variables and processes.* Izdat. Akad. Nauk. SSSR. 1960 (Russian) – Holden-Day, San Francisco. 1964.

Plessner, A. I. and Rohlin, V. A. [1]. Spectral theory of linear operators II. *Usp. Mat. Nauk.* **1** (1946), 71–191 (Russian) – *A.M.S. Transl. Ser.* 2. **62** (1967), 29–175.

Poincaré, H. [1]. *Les méthodes nouvelles de la mécanique céleste* vol. 3. Gauthier-Villars, Paris. 1899.

Pugh, C. and Shub, M. [1]. Ergodic elements of ergodic actions. *Compositio Mathematics* **23** (1971), 115–22.

Rényi, A. [1]. *Probability theory.* North-Holland, Amsterdam. 1970.

Riesz, F. [1]. Sur la théorie ergodique. *Comm. Math. Helv.* **17** (1945), 221–39.

Rohlin, V. A. [1]. On the fundamental ideas of measure theory. *Mat. Sborn.* (n.s.) **25** (1949), 107–150 (Russian) – *Amer. Math. Soc. Transl.* **71** (1952), 1–54.

Rohlin, V. A. [2]. Selected topics in the metric theory of dynamical systems. *Usp. Mat. Nauk.* (n.s.) **4** (1949), 57–128 (Russian) – *Amer. Math. Soc. Transl. Ser.* 2 **49** (1966), 171–240.

Rohlin, V. A. [3]. On the entropy of a metric automorphism. *Dokl. Akad. Nauk. SSSR.* **124** (1959) 980–3 (Russian).

Rohlin, V. A. [4]. Lectures on the entropy theory of transformations with invariant measure. *Usp. Mat. Nauk.* **22** (1967), 3–56 (Russian) – *Russian Math. Surveys* **22** (1967), 1–52.

Rohlin, V. A. [5]. Exàct endomorphisms of a Lebesgue space. *Izv. Akad. Nauk. SSSR. Ser. Mat.* **25** (1961), 499–530 (Russian) – *Amer. Math. Soc. Transl. Ser.* 2 **39** (1964), 1–36.

Rohlin, V. A. and Sinai, Ja. G. [1]. Construction and properties of invariant measurable partitions. *Dokl. Akad. Nauk. SSSR.* **141** (1961), 1038–41 (Russian) – *Sov. Math.* **2** (1961), 1611–14.

Rudolfer, S. M. and Wilkinson, K. M. [1]. A number-theoretic class of weak Bernoulli transformations. *Math. Systems Theory* **7** (1973), 14–24.

Ruelle, D. [1]. A measure associated with Axiom A. attractors. *Amer. J. Math.* **98** (1976), 619–54.

Schwartzman, S. [1]. Asymptotic cycles. *Ann. Math.* **66** (1957), 270–84.

Schwartzman, S. [2]. Global cross sections of compact dynamical systems. *Proc. Nat. Acad. Sci. U.S.A.* **48** (1962), 786–91.

Shannon, C. [1]. A mathematical theory of communication. *Bell System Tech. J.* **27** (1948), 379–423, 623–56.

Shields, P. [1]. *The theory of Bernoulli shifts.* University of Chicago Press. 1973.

Sinai, Ja. G. [1]. On the concept of entropy for dynamical systems. *Dokl. Akad. Nauk. SSSR.* **124** (1959), 768–71 (Russian).

Sinai, Ja. G. [2]. Classical dynamical systems with countably multiple Lebesgue spectrum. II. *Izv. Akad. Nauk. SSSR. Ser. Mat.* **30** (1966), 15–68 (Russian) – *Amer. Math. Soc. Transl. Ser.* 2 **68** (1968), 34–88.

Sinai, Ja. G. [3]. On the foundations of the ergodic hypothesis for a dynamical system of statistical mechanics. *Dokl. Akad. Nauk. SSSR.* **153** (1963), 1261–4 (Russian) – *Sov. Math. Dokl.* **4** (1963), 1818–22.

Sinai, Ja. G. [4]. On a weak-isomorphism of transformations with invariant measure. *Mat. Sborn.* **63** (1964), 23–42 (Russian) – *Amer. Math. Soc. Transl. Ser.* 2 **57** (1966), 123–43.

Sinai, Ja. G. [5]. Markov partitions and U-diffeomorphisms. *Func. Anal. Pril.* **2** (1968), 68–89 (Russian) – *Functional anal. and its applications* **2** (1968), 61–82.

Sinai, Ja. G. [6]. Gibbs measures in ergodic theory. *Usp. Mat. Nauk.* **27** (1972), 21–64 (Russian) – *Russian Math. Surveys* **166** (1972), 21–69.

Smale, S. [1]. Dynamical systems and the topological conjugacy problem for diffeomorphisms. *Proc. Int. Cong. Stockholm* 1962, pp. 490–496. Djursholm. 1963.

Smale, S. [2]. Differentiable dynamical systems. *Bull. Amer. Math. Soc.* **73** (1967), 747–817.

Smorodinsky, M. [1]. β-automorphisms are Bernoulli shifts. *Acta Math. Acad. Sci. Hung.* **24** (1973), 273–8.

Sternberg, S. [1]. *Celestial mechanics* vol. 1. Benjamin, New York. 1969.

von Neumann, J. [1]. Proof of the quasi-ergodic hypothesis. *Proc. Nat. Acad. Sci. U.S.A.* **18** (1932), 70–82.

von Neumann, J. [2]. Physical applications of the ergodic hypothesis. *Proc. Nat. Acad. Sci. U.S.A.* **18** (1932), 263–6.

von Neumann, J. [3]. Zur Operatoren methode in der klassischen mechanik. *Ann. Math.* **33** (1932), 587–642.

Walters, P. [1]. *Ergodic theory – introductory lectures.* Lecture notes in mathematics no. 458. Springer, Berlin. 1975.

Walters, P. [2]. Invariant measures and equilibrium states for some mappings which expand distances. *Trans. Amer. Math. Soc.* **236** (1978), 121–53.

Watson, G. N. [1]. *A treatise on the theory of Bessel functions.* Cambridge University Press, London. 1945.

Weiss, B. [1]. *Ergodic theory.* (Lectures.)

Weiss, B. [2]. Equivalence of measure-preserving transformations. (Lectures at Hebrew University, Jerusalem 1976.)

Weyl, H. [1]. Uber die Gleichverteilung von Zahlen mod 1. *Math. Ann.* **77** (1916), 313–52.

Wiener, N. [1]. The ergodic theorem. *Duke Math. J.* **5** (1939), 1–18.

Wilkinson, K. M. [1]. Ergodic properties of a class of piecewise linear transformations. (To appear.)

Zygmund, A. [1]. *Trigonometric series.* Cambridge University Press, London. 1959.

Further Literature

Brown, R. *Ergodic theory and topological dynamics*. Academic Press, New York. 1976.

Linnik, Yu. V. *Ergodic properties of algebraic fields. Ergebnisse der Mathematik und ihrer Grenzgebiete* vol. 45. Berlin. 1968.

Ornstein, D. An application of ergodic theory to probability theory. *Annals of Probability* **1**(1) (1973), 43–65.

Postnikov, A. G. Ergodic problems in the theory of congruences and of diophantine approximations. *American Mathematical Society* 1967. (Translation of *Proceedings of the Steklov Institute of Mathematics* **82** (1966).)

Weiss, B.'The isomorphism problem in ergodic theory, *Bull. Amer. Math. Soc.* **78** (1972), 668–84.

The following issues of journals are devoted entirely, or almost entirely, to ergodic theory:

Asterisque, Société mathématique de France **49, 50, 51** (1977).

Israel Journal of Mathematics **21** (nos. 2–3) (1975).

Russian Mathematical Surveys **22** (no. 5) (1967).

Zeitschrift für Wahrscheinlichkeitstheorie und verwandte Gebiete **13** (2) (1969).

The following is a selected list from the Springer Lecture Notes series with volume numbers and authors (editors):

(160) L. Sucheston (ed.); (206) D. Chillingworth (ed.); (214) M. Smorodinsky; (318) A. Beck (ed.); (334) F. Schweiger; (458) P. Walters; (468) A. Manning (ed.); (470) R. Bowen; (480) X. M. Fernique, J. P. Conze and J. Gani; (502) J. Galambos; (517) S. Glasner; (527) M. Denker, Ch. Grillenberger and K. Sigmund; (532) J. P. Conze and M. S. Keane (eds.); (668) J. C. Martin, N. G. Markley and W. Perrizo (ed.); (729) M. Denker and K. Jacobs.

Index

Printed in the United States
By Bookmasters